Rigging Equipment

Rigging Equipment

Maintenance and Safety Inspection Manual

Joseph A. MacDonald

New York   Chicago   San Francisco   Lisbon   London   Madrid
Mexico City   Milan   New Delhi   San Juan   Seoul
Singapore   Sydney   Toronto

The McGraw·Hill Companies

Cataloging-in-Publication Data is on file with the Library of Congress

Copyright © 2011 by The McGraw-Hill Companies, Inc. All rights reserved. Printed in the United States of America. Except as permitted under the United States Copyright Act of 1976, no part of this publication may be reproduced or distributed in any form or by any means, or stored in a data base or retrieval system, without the prior written permission of the publisher.

1 2 3 4 5 6 7 8 9 0   DOC/DOC   1 6 5 4 3 2 1 0

ISBN 978-0-07-171948-3
MHID 0-07-171948-2

Sponsoring Editor
Larry S. Hager

Editorial Supervisor
Stephen M. Smith

Production Supervisor
Richard C. Ruzycka

Acquisitions Coordinator
Michael Mulcahy

Project Manager
Sapna Rastogi, Glyph International

Copy Editor
Manish Tiwari, Glyph International

Proofreader
Upendra Prasad, Glyph International

Art Director, Cover
Jeff Weeks

Composition
Glyph International

Printed and bound by RR Donnelley.

McGraw-Hill books are available at special quantity discounts to use as premiums and sales promotions, or for use in corporate training programs. To contact a representative, please e-mail us at bulksales@mcgraw-hill.com.

This book is printed on acid-free paper.

Information contained in this work has been obtained by The McGraw-Hill Companies, Inc. ("McGraw-Hill") from sources believed to be reliable. However, neither McGraw-Hill nor its authors guarantee the accuracy or completeness of any information published herein, and neither McGraw-Hill nor its authors shall be responsible for any errors, omissions, or damages arising out of use of this information. This work is published with the understanding that McGraw-Hill and its authors are supplying information but are not attempting to render engineering or other professional services. If such services are required, the assistance of an appropriate professional should be sought.

ABOUT THE AUTHOR

Joseph A. MacDonald has more than 40 years of editorial experience, including 18 years with *Engineering News-Record* and 10 years with *Private Power Executive* magazine. He was Editor-in-Chief of McGraw-Hill's *Encyclopedia of U.S. Building and Construction Technology,* Editor of ENR's *Directory of Construction Information Resources,* and Co-Editor of *Handbook of Rigging,* now in its 5th Edition. Mr. MacDonald is a member of the Construction Writers Association, and a past president (two terms).

Contents

Preface xi
Cranes and Derricks in Construction Final Rule xv
Introduction xix

Part 1 Operations

Chapter 1. Operating Rules   3

   1.1 Rigging Hazards   3
   1.2 Regulations   7
   1.3 Standards (Consensus)   9
   1.4 Guidelines (Industry)   13

Chapter 2. Responsibilities   21

   2.1 Safety Organization   21
   2.2 Crane-Specific Responsibilities   22
   2.3 Project Responsibilities   25

Chapter 3. Procedures   33

   3.1 Planning and Preparation   33
   3.2 Inspections   37
   3.3 Standards (Consensus)   44

Part 2 Inspections

Chapter 4. Lifting Equipment   49

   4.1 Cranes and Derricks   49
   4.2 Forklifts   80
   4.3 Helicopters   88
   4.4 Jacks and Air Lifts   94

viii Contents

Chapter 5. Hoisting Equipment — 103

    5.1   Portable and Overhead Hoists — 103
    5.2   Material and Personnel Hoists — 114
    5.3   Winches and Drums — 121

Chapter 6. Rigging Systems — 129

    6.1   Wire Rope and Wire Rope Slings and Meshes — 129
    6.2   Alloy Steel Chain Slings — 135
    6.3   Natural Fiber Rope Slings and Hitches — 137
    6.4   Synthetic Fiber Ropes and Web Slings — 140
    6.5   Standards (Consensus) — 143

Chapter 7. Rigging Hardware — 145

    7.1   Overhead Lifting Systems — 145
    7.2   Attachments, Fittings, and Connections — 148
    7.3   Hooks and Shackles — 153
    7.4   Sheaves, Blocks, and Tackles — 155
    7.5   Below-the-Hook Lifting Devices — 156
    7.6   Standards (Consensus) — 157

Chapter 8. Scaffolding Systems — 159

    8.1   Guidelines for Setup, Use, and Maintenance — 159
    8.2   Suspended Scaffolds — 165
    8.3   Stationary Scaffolds — 166
    8.4   Mobile Scaffolds — 170
    8.5   Aerial Lifts (Platforms) — 171
    8.6   Construction Scaffolds — 172
    8.7   Scaffold Access — 181
    8.8   Guardrail Systems, Canopies, and Covers — 186
    8.9   Regulations — 191
    8.10 Standards (Consensus) — 192

Chapter 9. Fall Protection — 195

    9.1   Fall Arrest Systems — 195
    9.2   Personal Fall Protection Arrest Systems — 214
    9.3   Safety Net Systems — 227
    9.4   Regulations — 232
    9.5   Standards (Consensus) — 232

Chapter 10. Ladders, Stairways, and Ramps — 235

    10.1 General Industry Ladders — 241
    10.2 Fixed Ladders — 243
    10.3 Portable Ladders — 246

Contents   ix

10.4	Mobile Ladder Stands	252
10.5	Construction Site Ladders	253
10.6	Fixed Stairways and Ramps	263
10.7	Walkways and Elevated Platforms	266
10.8	Guarding Roofs, Floors, Stairs, and Other Openings	268
10.9	Standards (Consensus)	272

Chapter 11. Tools and Machinery   273

11.1	Hand and Portable Power Tools	273
11.2	Welding and Cutting (Hot Work)	281
11.3	Compressors and Compressed Air	284
11.4	Gas Cylinders	285
11.5	Machine Guarding and Lockout/Tagout	287
11.6	Standards (Consensus)	289

Chapter 12. Electrical Systems   291

12.1	Electrical Safety	291
12.2	Safe Work Practices	298
12.3	Hazardous Energy Control	301
12.4	Standards (Consensus)	304

Chapter 13. Worker Protection   307

13.1	Health and Safety	308
13.2	Medical Services/First Aid	309
13.3	Personal Protective Equipment and Clothing	311
13.4	Resources	318

Part 3   Protection

Chapter 14. Protection Systems   323

14.1	Fencing	329
14.2	Lighting	330
14.3	Signage	332
14.4	Barricades	333
14.5	Standards (Consensus)	334

Chapter 15. Protective Procedures   335

15.1	Fire Safety	335
15.2	Hazardous and Toxic Materials	340
15.3	Theft and Vandalism	343
15.4	Worksite and Equipment Security	347
15.5	Unsafe Site Conditions	349
15.6	Standards (Consensus)	349

Appendix A Bibliography 351
Appendix B Resources 359
Appendix C Regulations and Standards 367
Index 381

Preface

Many types of cranes, hoists, and rigging devices are used today, for both construction and industrial applications, to lift and move materials—with the project contractor or facility manager responsible for maintaining a safe workplace through implementation of equipment safety rules and guidelines. These apply to all operations that involve the use of cranes, hoists, and rigging devices, regardless of whether they are on the worksite, or installed in or attached to buildings that are used in

- Erection, installation, or dismantling of materials and equipment
- Renovation, maintenance, or demolition of buildings or structures

Equipment safety rules and guidelines apply to all employees, supplemental labor, and subcontractor personnel who use such devices. And it cannot be overemphasized that, above all, only qualified and licensed individuals should operate such devices.

Rigging Equipment: Maintenance and Safety Inspection Manual is a companion volume to McGraw-Hill's industry-acknowledged standard reference—*Handbook of Rigging: Lifting, Hoisting, and Scaffolding for Construction and Industrial Operations,* now in its 5th Edition. This new inspection manual is intended to help rigging contractors, facility managers, and equipment operators in maintaining and operating rigging equipment safely.

The manual includes regulations, standards, guidelines, and recommendations applicable to critical lifts; provides maintenance and safety inspection checklists for rigging equipment, components, and systems; and addresses the training, planning, and documentation required to perform critical lifts. The safe rigging practices recommended in this manual of necessity are framed in general terms to accommodate the many variations in rigging practices and the different ways that rigging is used.

Checklists included in this manual were developed from copyright-free government publications, general consensus standards, and industry association/equipment manufacturers' guidelines.

Federal Regulations

OSHA—Occupational Safety & Health Administration

MSHA—Mine Safety & Health Administration

NIOSH—National Institute for Occupational Safety & Health

Consensus Standards

ANSI—American National Standards Institute

ASME—American Society of Mechanical Engineers

ASSE—American Society of Safety Engineers

ASTM—ASTM International

NFPA—National Fire Protection Association

SAE—Society of Automotive Engineers

Industry Guidelines

ALI—American Ladder Institute

AWS—American Welding Society

CI—Cordage Institute

CMAA—Crane Manufacturers Association of America

HMI—Hoist Manufacturers Institute

MHIA—Material Handing Industry of America

NEMA—National Electrical Manufacturers Association

PCSA—Power Crane & Shovel Association

SIA—Scaffolding Industry Association

SSFI—Scaffolding, Shoring, and Forming Institute

UL—Underwriters Laboratory

WRTB—Wire Rope Technical Board

Note: Because guidelines and recommendations are only advisory in nature, they must be supplemented by proper training—and strict observance.

Maintenance and Inspection

Rigging equipment, like any machine, requires regular maintenance and inspection. Lack of maintenance will increase the likelihood of a component failure, which could be catastrophic, resulting in property damage, serious injuries, or death, while properly maintained equipment should provide trouble-free operation for many years.

Regular inspection of all equipment and systems used in rigging operations not only will greatly reduce the risk of any failures, but also is required by OSHA regulations, along with the maintenance of

- Data base—All hoisting and lifting equipment and rigging systems (project hoisting and rigging inspector)
- Equipment records—All hoisting and lifting equipment and rigging systems (equipment manager)
- Proper documentation—All services rendered on that equipment (project hoisting and rigging inspector)

Rigging equipment inspections must include initial, pre-use, periodic, and third party, with two persons sharing the responsibilities—one to ensure performance of inspections and the other to perform the inspection. Normally, such inspections require use of lengthy checklists that identify mechanical components and maintenance schedules. Too often, however, checklists provide inadequate description or explanations pertinent to the relationship between components and the equipment's overall function. This manual attempts to correct this deficiency:

- Providing background information on hoisting and lifting principles of rigging equipment
- Discussing relationships between various components of equipment and their lifting capacity
- Presenting insurance company recommendations for conducting proper maintenance inspections
- Publishing industry maintenance and safety inspection checklists

Although most of the inspection checklists included in this manual are self-explanatory, users must recognize that due to increasing applications of developing technology in the design and manufacture of cranes, rigging equipment inspectors require a better understanding of crane operations and their basic lifting principles, and they must keep abreast of related developments in today's hoisting and lifting environment.

Since cranes affect a large segment of work at any construction site, crane inspection by the project safety manager must include a survey of the entire operation, including how the crane will be operating and how other crafts will be affected by working with and around the crane. Observing crane operations before an inspection, or asking questions about how it will or has been operating, can indicate possible problem areas that may need a closer review during the inspection process.

Joseph A. MacDonald

Cranes and Derricks in Construction Final Rule

Fact Sheet
July 28, 2010

The U.S. Department of Labor's Occupational Safety and Health Administration (OSHA) released a historic new standard on July 28, 2010, addressing the use of cranes and derricks in construction and replacing a decades old standard. The significant number of fatalities associated with the use of cranes and derricks in construction and the considerable technological advances in equipment since the publication of the old rule, issued in 1971, led the Labor Department to undertake this rulemaking.

In 1998, OSHA's expert Advisory Committee on Construction Safety and Health (ACCSH) established a workgroup to develop recommended changes to the current standard for cranes and derricks. In December 1999, ACCSH recommended that the Agency use negotiated rule making to develop the rule. The Cranes and Derricks Negotiated Rulemaking Committee (C-DAC) was convened in July 2003 and reached consensus on its draft document in July 2004. In 2006, ACCSH recommended that OSHA use the C-DAC consensus document as a basis for OSHA's proposed rule, which was published in 2008. Public hearings were held in March 2009, and the public comment period on those proceedings closed in June 2009.

- The rule becomes effective 90 days after August 9, 2010, the date the final rule will be published in the Federal Register. Certain provisions have delayed effective dates ranging from 1 to 4 years.
- A copy of the regulatory text is available at: http://www.osha.gov/doc/cranesreg.pdf.
- Until the date of publication, the full rule, including the preamble, can be found at: http://www.ofr.gov/inspection.aspx. After publication, the rule can be found at the Federal Register or at www.osha.gov.
- This new standard will comprehensively address key hazards related to cranes and derricks on construction worksites, including the four

main causes of worker death and injury: electrocution, crushed by parts of the equipment, struck-by the equipment/load, and falls.
- Significant requirements in this new rule include a pre-erection inspection of tower crane parts; use of synthetic slings in accordance with the manufacturer's instructions during assembly/disassembly work; assessment of ground conditions; qualification or certification of crane operators; and procedures for working in the vicinity of power lines.
- This final standard is expected to prevent 22 fatalities and 175 non-fatal injuries each year.
- Several provisions have been modified from the proposed rule. For example:
  - Employers must comply with local and state operator licensing requirements which meet the minimum criteria specified in § 1926.1427.
  - Employers must pay for certification or qualification of their currently uncertified or unqualified operators.
  - Written certification tests may be administered in any language understood by the operator candidate.
  - When employers with employees qualified for power transmission and distribution are working in accordance with the power transmission and distribution standard (§ 1910.269), that employer will be considered in compliance with this final rule's requirements for working around power lines.
  - Employers must use a qualified rigger for rigging operations during assembly/disassembly.
  - Employers must perform a pre-erection inspection of tower cranes.
- This final rule requires operators of most types of cranes to be qualified or certified under one of the options set forth in § 1926.1427. Employers have up to 4 years to ensure that their operators are qualified or certified, unless they are operating in a state or city that has operator requirements.
- If a city or state has its own licensing or certification program, OSHA mandates compliance with that city or state's requirements only if they meet the minimum criteria set forth in this rule at § 1926.1427.
- The certification requirements in the final rule are designed to work in conjunction with state and local laws.
- This final rule clarifies that employers must pay for all training required by the final rule and for certification of equipment operators employed as of the effective date of the rule.

- State Plans must issue job safety and health standards that are "at least as effective as" comparable federal standards within 6 months of federal issuance. State Plans also have the option to promulgate more stringent standards or standards covering hazards not addressed by federal standards.
- OSHA has additional compliance assistance material available.

Introduction

Government and industry accident statistics indicate that project management and engineers face a great challenge to drastically reduce rigging- and scaffolding-related accidents—and to ensure that all workers, operators, and foremen are well trained in the basics of safe lifting and hoisting practices. Past inspection evidence shows particular crane hazards—falls, being struck by objects, electrocution, crane tip-over, and being caught in or between machinery—are the leading causes of accidents where cranes are used on construction sites and for industrial operations.

From the use of computers to hydraulic systems, crane technology has changed considerably over the past 30 years, and the American Society of Mechanical Engineers (ANSI's standards-developing organization for B-30 standards) continually updates the standards for crane manufacturing, operational procedures, inspection requirements, and operator qualifications. Conflicts between OSHA's heavy reliance on a late-1960s version of the ANSI B-30.5 standard for crawler cranes and the more up-to-date ANSI B-30.5 standard, however, have continually posed dilemmas for construction firms. As a result, the rash of crane-tipping accidents that occurred in 2008 has triggered crane manufacturers, construction companies, and others to press OSHA to update its crane rule—not only as genuine concern for safety, but also for economic self-interest.

Losses stemming from a crane failure do not stop with the incident itself, however, even if people are not injured or killed. According to casualty insurance providers, after a crane accident productivity drops dramatically with everyone focused on what has occurred. Insurance statistics prove that in the aftermath of a major crane failure, incident rates increase because instead of paying attention to their safety, workers are preoccupied with the accident. According to insurance records, supervisors are besieged, not only with the consequences of the disaster but also with smaller problems—such as lawsuits, bad press, higher insurance premiums, or the loss of insurance coverage altogether.

The first person who makes a decision about a crane is the estimator, and then the superintendent and project manager. Too many times, though, they don't have sufficient crane knowledge to make the right decisions. Crane safety specialists point out that "whenever there is a crane on site, everyone involved should be afraid, because it takes so many people executing correctly for it to work."

A crane is by far the most dangerous and costly piece of equipment on a worksite—with three components to crane safety:

- Crane design
- Operating environment
- Operator control

According to crane safety specialists, the two most common crane management problems are not setting up the crane properly and supervisor intimidation of operators. Too often, supervisors ask operators to overload the crane. Earlier standards granted operators the right to refuse a lift they judged unsafe, but if an operator refuses to do a job today, the supervisor will find some else to do it.

From a safety perspective, the most fundamental change in crane technology over the past three decades has been the shift from mechanical cranes with brakes and friction that the operator could "feel" to hydraulic cranes. Hydraulic cranes today require that operators be properly trained to understand the capability of the crane they are using.

The growth in the number and complexity of attachments is a second key development in the evolution of crane design over the past 30 years. While customers demand lighter cranes with greater reach and lifting capacity, the new technology also requires more of operators and managers if these cranes are to be used safely.

Safety specialists point out that "... while the physical demands of crane operation have become easier, the mental requirements have grown. Cranes now operate with a joystick, like video games. And, today's cranes have on-board computers," that they say, "require an additional amount of training as opposed to relying on the skill and judgment of the operator."

With the exception of a few states and municipalities, no special license or certification is required of crane operators. Some safety experts believe that because today's cranes are so sophisticated, all operators should be certified. They point out that the failure to spell out operator qualifications is an unacceptable gap in the current OSHA crane rules—and too many contractors do a very poor job of managing cranes.

Thus, this easy-to-use manual of equipment maintenance and safety inspections was developed for use by equipment operators and inspectors, as well as safety and health managers, and project supervisors.

Chapters cover

- Operator qualifications and operating procedures
- Rigging requirements
- Planning and preparation before performing rigging tasks
- Proper use of personal safety and protective equipment
- Maintenance schedules, care, and safe operation of equipment
- Inspection checklists for rigging equipment before, during, and after use
- Testing, certification, and registration of rigging equipment
- Recordkeeping of preventative maintenance program based on equipment manufacturers' recommendations.

Checklists exist because safety is paramount on any construction job or facility project. There are many dangers present on a construction or facility worksites, that could lead to disasters that are best avoided. It is, therefore, of the utmost importance that the principal contractor for a certain construction job make sure that the usual hazards are watched out for. It is a principal contractor's obligation to keep aware of the many risks he or she must avert before starting and also during construction work.

Construction safety checklists definitely come in handy when just about to start on a construction job. They significantly lessen the chances of the principal contractor to not identify risks in the construction site, especially those that are most frequently identified in other construction jobs. Furthermore, they help to more easily check whether a construction site complies with workplace health and safety regulations.

Given all this, it is good practice to build construction safety checklists and continually add on to them as the principal contractor identifies more potential hazards:

Site inspection—Walk around the site and check for any small structure present that wasn't included in the plans and may need to be demolished, or a depression that could possibly indicate the presence of something that has been buried—this may even lead to discovery of an underground tank that would require EPA involvement. You may also choose to hire an independent contractor to do the site inspection for you.

Employee site orientation—Prior to starting construction work, the workers must be made aware of the unique risks and hazards present in the site. Knowing what to look out for greatly helps to avert disasters.

Worksite fenced off—Placing a fence around the site keeps outsiders from getting in and exposing themselves to potential dangers. Also, these trespassers may end up taking equipment or materials.

Protective equipment—After having assessed the risks in the site, the principal contractor must make sure that all protective equipment has been provided and that all workers are properly making use of such equipment.

First aid—Verify that first-aid facilities and first-aid equipment are readily available and accessible in the worksite. First-aid response must be quick and done properly.

Construction signs—Place signs informing the general public of the potential hazards of being close to the site.

Tools and equipment—Testing tools and equipment to verify that they are working properly is another measure that needs to be taken just before construction work starts. This must also be applied to all electrical leads or outlets.

The checklists in this manual are merely to start off with and build up on. They may be a good start, though, for any new contractor intending to prepare reliable construction safety checklists.

Caveat: McGraw-Hill's *Rigging Equipment: Maintenance and Safety Inspection Manual* is not intended to be a legal interpretation of the provisions of the Occupational Safety & Health Administration Regulations (law), or Industry Consensus Standards, or to place any additional requirements on employers or employees.

Rigging Equipment

Part 1

Operations

Chapter 1

Operating Rules

According to safety experts, 50 percent of all crane accidents are attributed to improper setup. The critical points of safe operations, they point out, include:

- Setting the crane up on firm, level ground
- Using hardwood mats or blocking when necessary under outriggers
- Avoiding setup in hazardous areas
- Using ropes or barricades to prevent entry into the lift area
- Inspecting the setup and the crane

1.1 Rigging Hazards

The legal requirements for those in authority or in any supervisory capacity on Construction Worksites or in Industrial Facilities include "properly training for competency" the individuals and technicians who are under their care as employees, contractors, engineers, and visitors subject to these hazardous exposures. The liability for not notifying and training all personnel is extremely high.

Injuries, incidents, deaths, and major law suits proceed directly from the lack of implementation of Health and Safety Programs and Competency Training.

The following selected list of typical hazards and hazardous exposures to Health and Safety encountered on Construction Worksites and in Industrial Facilities is NOT definitive, but rather it is a beginning of representing areas that everyone in Construction Work and Facilities Operations needs to be cautioned about.

4   Operations

The lists in this manual, pertinent to a particular Construction Worksite or Industrial Facility, should be upgraded regularly and added to, as well as tailored specifically to each Worksite or Facility.

Health hazards
- Fatigue due to extended hours and overtime
- Health from aerosol sprays in close enclosures
- Not applying ergonomic techniques in all tasks
- Lack of proper eye protection, face masks, goggles, safety glasses
- Using pneumatic tools without the proper training by a competent professional
- Wearing improper footwear in shops, on location
- Exposure to impact and high-decibel noise
- Exposure to noise when proper noise attenuation has not been applied to walls, ceilings, and equipment
- Lack of and reasonable care of proper personal protective equipment that is mandated by Federal Regulations
- Lack of proper first-aid practices
- Incorrect lifting and moving equipment
- Lack of proper hand protection, gloves, creams
- Proper bodily protection in every way
- Using pneumatic nailers and other pneumatic devices without full personal protection, hands, face shields, body shields
- Lack of foot protection, work shoes, steel-toed boots
- Lack of a federally mandated Hearing Conservation Program and proper personal equipment, earplugs or muffs
- Not using proper hearing protection and not insulating loud and noisy equipment in the work environment
- Personnel called upon to do too much in short period of times, under stress, pressure, fatigue

Indoor air quality and ventilation hazards
- Lack of proper ventilation practices and exhaust equipment, (fans, booths, circulation of fresh air)
- Not regularly scheduling high-efficiency particulate air (HEPA) vacuuming work areas and washing down surfaces

- Contamination in ducting, showers, sinks, fan outlets, and air intake units, and from bacteria from microbes
- Carbon monoxide, closed areas, vehicle emissions within garages, studios, enclosed areas of any type
- Not having properly mandated, engineered ventilation in compliance with codes and standards
- Lack of proper storage and use of chemicals in non-ventilated working environments
- Solvents, used without proper personal protective equipment, lack of ventilation, fire hazards, stored incorrectly
- Toxic fumes, vapors, or emissions

Tripping, slipping, and falling hazards

- Poor floor surfaces, lack of proper maintenance, slippery surfaces, tripping objects
- Not properly abrading steps and tread surfaces
- Not properly cleaning floors and stairs and other surfaces, including walls, grids, and behind lock rails

Chemical exposure hazards

- Use, storage, lack of facility protection, fire, lack of proper personal protective devices, improper waste storage and disposal
- Lack of training personnel in the use of Material Safety Data Sheets (MSDS)
- Cleaning metal surfaces, skin, eyes, vapors, fire, burns
- Lack of knowledge and misuse of MSDS
- MSDS having misinformation
- MSDS that are not updated with latest findings

Electrical hazards

- Working near high-power lines, especially with lifts and cranes
- Open electrical circuit boxes, connections, wires
- Improper lighting installations, C-clamp breakage, abraded wires, exposed wires, lack of fusing, open lighting cabinets
- Electrical devices and electricity, cables, equipment
- Electrical systems in noncompliance with required National Electrical Code (NEC) regulations

- Lack of grounding equipment, Ground Fault Interrupter Circuits (GFICs), etc.
- Non-maintained lighting instruments

Fire hazards
- Lack of proper fire prevention practices
- Improper application of National Fire Protection Association (NFPA) and Standard of Care Fire Protection Methods and Procedures
- Blocked fire hoses and extinguishers
- Storing too close to sprinkler heads
- Not closing and propping open fire doors
- Smoking around areas where aerosol sprays are used
- Improper storage of flammable or combustible materials (or both)
- Flammable liquids not used correctly, not stored properly, incorrect containers, noncompliance with industry and NFPA standards, spills, improper ventilation, smoking around these liquids, incorrectly marked and non-marked containers, vapors
- Lack of training with fire extinguishers
- Hanging items from sprinkler pipes and heads

Equipment hazards
- Lack of preventive maintenance of motors
- Use of broken and worn tools
- Lack of safeties on overhead equipment
- Incorrect application of motors
- Lack of proper guards on power equipment
- Lack of redundancy on overhead equipment

Rigging hazards
- Rope: its use, lack of maintenance and untimely replacement, wrong types, fibers, burns, breaking, wrong knots, cuts, UV damage, corrosion, rotting, drying out
- Defective rigging equipment
- Lack of preventive maintenance of rigging equipment
- Lack of training in safe rigging practices

Fall hazards

- FAILURE to mark or barricade the edges, stairs, ramps, and floor openings or working area
- FAILURE to follow Occupational Safety and Health Administration (OSHA) regulations applied to the use of openings or edges of elevated platforms, ramps, trusses, and lifts

1.2 Regulations

General rigging safety

29 CFR 1910 Subpart B   Adoption and Extension of Established Federal Standards
   .12   Construction work

29 CFR 1910. Subpart F   Powered Platforms, Manlifts, and Vehicle-Mounted Work Platforms
   .66   Powered platforms for building maintenance
   .67   Vehicle-mounted electing and rotating work platforms
   .68   Manlifts

29 CFR 1910 Subpart I   Personal Protective Equipment
   .133   Eye and face protection
   .134   Respiratory protection
   .135   Head protection
   .136   Occupational foot protection
   .137   Electric protective devices
   .139   Hand protection

29 CFR 1910 Subpart N   Materials Handling and Storage
   .178   Powered industrial trucks
   .179   Overhead and gantry cranes
   .189   Crawler, locomotive and truck cranes
   .181   Derricks
   .182   Helicopters
   .183   Slings

Construction rigging safety

29 CFR 1926 Subpart C   General Safety and Health Provisions
   .26   Personal protective equipment

29 CFR 1926 Subpart E   Personal Protective and Lifesaving Equipment
   .95   Criteria for personal protective equipment
   .96   Occupational foot protection

8   Operations

.100 Head protection
.101 Hearing protection
.102 Eye and face protection
.103 Respiratory protection
.104 Safety belts, lifelines, and lanyards
.105 Safety nets

29 CFR 1926 Subpart H   Materials Handling Storage, Use, and Disposal

.250 General requirements for storage
.251 Rigging equipment for material handling

29 CFR Subpart L   Scaffolds

.451 General requirements
.452 Requirements applicable to specific types of scaffolds
.453 Aerial lifts
Appendix B   Criteria for determining the feasibility of providing safe access and fall protection for scaffold erectors and dismantlers

29 CFR 1926 Subpart M   Fall Protection

.502 Fall protection systems criteria and practices

29 CFR 1926 Subpart N   Cranes, Derricks, Hoists, Elevators, and Conveyors

.550 Cranes and derricks
.551 Helicopters
.552 Material hoists, personnel hoists, and elevators
.553 Base-mounted drum hoists
.554 Overhead hoists
.556 Aerial lifts

29 CFR 1926 Subpart O   Motor Vehicles and Mechanized Equipment

.600 Equipment
.601 Motor vehicles
.602 Material handling equipment

29 CFR 1926 Subpart R   Steel Erection

.753 Hoisting and rigging
.754 Structural steel assembly
.755 Column anchorage
.756 Beams and columns
.756 Open web steel joists
.759 Falling object protection
.760 Fall protection

29 CFR 1926 Subpart W   Rollover Protective Structures (ROPS); Overhead Protection

.1000 Rollover protective structures for material handling equipment
.1001 Minimum performance criteria for rollover protective structures for loaders
.1002 Protective frames (ROPS) for wheel-type agricultural and industrial tractors used in construction

29 CFR Subpart X   Ladders
.1052 Stairways
.1053 Ladders

1.3 Standards (Consensus)

ANSI	American National Standards for Construction and Demolition Operations
ASME	American Society of Mechanical Engineers
ASSE	American Society of Safety Engineers
CMAA	Crane Manufacturers Association of America
PCSA	Power Crane and Shovel Association

Safety and health program

ASSE/SAFE A10.33	Safety and Health Program Requirements for Multiemployer Projects
	American National Standard Construction and Demolition Operations
ASSE/SAFE A10.38	Basic Elements of an Employer's Program to Provide a Safe and Healthful Work Environment
	American National Standard Construction and Demolition Operations
ASSE/SAFE A10.39	Construction Safety and Health Audit Program
	American National Standard for Construction and Demolition Operations
ASSE/SAFE A10.42	Safety Requirements for Rigging Qualifications and Responsibilities
	American National Standard for Construction and Demolition Operations

ASSE/SAFE Z490.1	Accepted Practices in Safety, Health and Environmental Training
ASSE/SAFE CONSTRUCTION	Construction Safety Management and Engineering—Comprehensive Safety Resource Covering Program Essentials, Best Practices, Legal and Regulatory Requirements and Real World Guidance on Technical Issues

Personal protective equipment and clothing

ANSI Z87	Eye and Face Protection
ANSI Z87.1 (2003)	Personal Eye and Face Protection Devices
ANSI Z88	Respiratory protection
ANSI S12	Hearing Protection
ANSI/ISEA Z89.1-2009	Industrial Head Protection
ANSI/ISEA 105-2005	Hand Protection Selection Criteria
ANSI Z41-1999	Protective Footwear
ANSI Z41	User Guide for Protective Footwear
ANSI/ISEA 107-2007	High-Visibility Safety Apparel and Headwear
US TAG—ISO TC 94/SC13	Protective Clothing

Derricks, cranes and hoists

ASSE/SAFE CRANE HAZARDS	Crane Hazards and Their Prevention (50 Categories of Crane Hazards, Including Power Line Contact), Types of Upset and Pinch Points and Nip Points
ASSE/SAFE A10.28	Safety Requirements for Work Platforms Suspended from Cranes or Derricks—American National Standard for Construction and Demolition Operations
ASSE/SAFE A10.42	Safety Requirements for Rigging Qualifications and Responsibilities—American National Standard for Construction and Demolition Operations

ASME B30.6	Derricks
ASME B30.5	Overhead and Gantry Cranes (Top-Running Bridge, Single or Multiple Girder, Top-Running Trolley Hoist)
ASME B30.16	Overhead Hoists (Underhung)
ASME B30.17	Overhead and Gantry Cranes (Top-Running Bridge, Single Girder, Underhung Hoist)
ASME B30.11	Monorail Systems and Underhung Cranes
CMAA No.70	Electric Overhead Traveling Cranes—Specifications
CMAA No. 74	Top- and Under-Running Single Girder, Electric Overhead Traveling

Motor vehicles

ASSE/SAFE Z15.1	Safe Practices for Motor Vehicle Operations

Hoists (wire rope and chain)

ANSI/ASME/HST-3M	Performance Standard for Manually Lever-Operated Chain Hoists
ANSI/ASTM/HST-4M	Performance Standard for Electric Wire Rope Hoists
ANSI/ASME HST-5M	Performance Standard for Air Chain Hoists
ANSI/ASME HST-6M	Performance Standard for Air Wire Rope Hoists

Personnel and materials hoists

ASSE/SAFE A10.4	Personnel Hoists and Employee Elevators on Construction and Demolition Sites—American National Standard for Construction and Demolition Operations

Fall protection

ASSE/SAFE FALL PROTECTION	Introduction to Fall Protection—3rd Edition—Identification of Specific Walking and Working Surface Hazards, Including Slips and Trips, Stairways and Ramps, Ladders, Scaffolds, and Roofs

Operations

ASSE/SAFE Z359.2	Minimum Requirements for a Comprehensive Managed Fall Protection Program
ASSE/SAFE Z359.3	Safety Requirements for Positioning and Travel Restraint Systems
ASSE/SAFE Z359.4	Safety Requirements for Assisted-Rescue and Self-Rescue Systems, Subsystems and Components
ASSE/SAFE A10.24	Roofing—Safety Requirements for Low-Sloped Roofs—American National Standard for Construction and Demolition Operations

Personnel and debris nets

ASSE/SAFE A10.11	Safety Requirements for Personnel and Debris Nets—American National Standard for Construction and Demolition Operations
ASSE/SAFE A10.32	Fall Protection Systems—American National Standard for Construction and Demolition Operations

Walking/working surfaces

ASSE/SAFE A1264.1	Safety Requirements for Workplace Walking/Working Surfaces and Their Access; Workplace, Floor, Wall, and Roof Openings; Stairs and Guardrails Systems
ASSE/SAFE A10.18	Safety Requirements for Temporary Roof and Floor Holes, Wall Openings, Stairways, and Other Unprotected Edges in Construction and Demolition Operations—American National Standard for Construction and Demolition Operations
ASSE/SAFE Z117.1	Safety Requirements for Confined Spaces
ASSE/SAFE A1264.2	Provision of Slip Resistance on Walking/Working Surfaces

Energy lockout/tagout

ASSE/SAFE Z244.1	Control of Hazardous Energy Lockout/Tagout and Alternative Methods

Steel erection

ASSE/SAFE A10.13　　Safety Requirements for Steel Erection—American National Standard for Construction and Demolition Operations

Powder-actuated tools

ASSE/SAFE A10.3　　Safety Requirements for Powder-Actuated Fastening Systems—American National Standard for Construction and Demolition Operations

Site protection

ASSE/SAFE A10.34　　Protection of the Public on or Adjacent to Construction Sites—American National Standard for Construction and Demolition Operations

Demolition operations

ASSE/SAFE A10.6　　Safety Requirements for Demolition Operations—American National Standard for Construction and Demolition Operations

Masonry and concrete work

ASSE/SAFE A10.9　　Safety Requirements for Masonry and Concrete Work—American National Standard for Construction and Demolition Operations

Welding and cutting

ANSI Z49　　Safety in Welding and Cutting

1.4　Guidelines (Industry)

- Know the hoist lifting capacity.
- Always know the weight of the object you are lifting.
- Never exceed the working load limit.
- Train operators on proper rigging techniques as well as hoist operation.
- Have rigging handbooks and proper equipment available for use.

- When rigging, make sure the load hook and upper suspension form a straight line.
- Chain or body of the hoist should never come in contact with the load.
- Never tip-load hooks.
- Always use a sling or lifting device to rig around loads and use engineered lift points for attachment.
- Never work under suspended loads or lift loads over people.
- Never lift people with a hoist.
- When leaving the hoist unattended, land any attached loads.
- When the job is complete, place the hoist and hook in a location that will not interfere with the movement of people or materials.

Source: Industrial Training International, Woodland, WA.

SAFETY POLICIES AND PROCEDURES	YES	NO	N/A
1. Is there a written, unit-specific policy manual that includes information pertaining to safety policies and procedures?	☐	☐	☐
2. Is there an active unit safety committee or designated unit safety contact?	☐	☐	☐
3. Are safety meetings held on a regular basis?	☐	☐	☐
4. Are safety bulletin board, web site, and/or other means being used to disseminate safety information throughout the unit?	☐	☐	☐
5. Is every employee given a basic safety orientation by their immediate supervisor (or designee) on their first day of the job?	☐	☐	☐
6. Have all employees completed a Slip, Trip, and Fall Prevention course given by a qualified supervisor?	☐	☐	☐
7. Is at least one person at the worksite trained in standard first-aid/CPR?	☐	☐	☐
8. Have employees, who are designated as first-aid responders or working in other positions where they might reasonably be expected to encounter blood or other bodily fluids, completed a Blood Borne Pathogen Training course presented by a qualified supervisor?	☐	☐	☐

9. Has other safety training, needed to perform each job within the unit, been identified, provided, and documented as required? ☐ ☐ ☐

10. Are there suitable first-aid kits and fire extinguishers available in all project or facility vehicles and each work area? ☐ ☐ ☐

11. Are first-aid kits readily visible, accessible, regularly inspected, and maintained? ☐ ☐ ☐

12. Are emergency telephone numbers posted on all first-aid kits and telephones? ☐ ☐ ☐

13. Has a written Emergency Response Plan been completed for and disseminated throughout the unit? ☐ ☐ ☐

14. Have supervisors of employees who may operate or service equipment or perform tasks (or both) that could result in serious injuries, if safe work practices are not identified and followed, completed a Job Hazard Analysis course presented by a qualified supervisor? ☐ ☐ ☐

15. Have personnel elected or appointed to serve on unit safety committees; unit managers, Supervisors, and other personnel who may be involved in the accident reporting or remediation process attended a Reporting, Reviewing, and Reducing Accidents and Incidents course presented by a qualified supervisor? ☐ ☐ ☐

FACILITIES

YES   NO   N/A

Exits

1. Are exit signs provided with the word "EXIT" in lettering at least 6 in high and the stroke of the lettering at least 3/4 in wide? ☐ ☐ ☐

2. Are exit doors side-hinged? ☐ ☐ ☐

3. Are all exits kept free of obstructions and unlocked? ☐ ☐ ☐

4. Are at least two means of egress provided from elevated platforms, pits, or rooms where the absence of a second exit would increase the risk of injury from hot, poisonous, corrosive, suffocating, flammable, or explosive substances? ☐ ☐ ☐

Operations

5. Are there sufficient exits to permit prompt escape in case of emergency? ☐ ☐ ☐
6. Are the number of exits from each floor of a building and the number of exits from the building itself appropriate for the building occupancy load? ☐ ☐ ☐
7. When workers must exit through glass doors, storm doors, etc., are the doors fully tempered and meeting safety requirements for human impact? ☐ ☐ ☐

Exit Doors

8. Are doors required to serve as exits designed and constructed so that the way of exit travel is obvious and direct? ☐ ☐ ☐
9. Are windows (that could be mistaken for exit doors) made inaccessible by barriers or railing? ☐ ☐ ☐
10. Are exit doors able to open from the direction of exit travel without the use of a key or any special knowledge or effort?
11. Are revolving, sliding, or overhead doors PROHIBITED from serving as exit doors? ☐ ☐ ☐
12. When panic hardware is installed on a required exit door, will it allow the door to open by applying a force of 15 lb (2.268 kg) or less in the direction of the exit traffic? ☐ ☐ ☐
13. Do doors on cold-storage rooms have an inside release mechanism that will release the latch and open the door even if it is padlocked or otherwise locked on the outside? ☐ ☐ ☐
14. Where exit doors open directly onto a street, alley, or other area where vehicles may be operated, are adequate barriers and warnings provided to prevent employees from stepping directly into the path of traffic? ☐ ☐ ☐
15. Are doors that swing in both directions between rooms, in which there is frequent traffic, provided with viewing panels in each door? ☐ ☐ ☐

ENVIRONMENTAL CONTROLS   YES NO N/A

1. Are all work areas properly illuminated? ☐ ☐ ☐
2. Are employees instructed in proper first-aid and other emergency procedures? ☐ ☐ ☐

3. Are hazardous substances, blood and other potentially infectious materials, which may cause harm by inhalation, ingestion, or skin absorption or contact, identified?

4. Are employees aware of the hazards involved with various chemicals they may be exposed to in the work environment, such as ammonia, chlorine, epoxies, caustics, etc.?

5. Is employee exposure to chemicals in the workplace kept within acceptable levels?

6. Can a less harmful method or product be used?

7. Is the work area ventilation system appropriate for the work performed?

8. Are spray painting operations performed in spray rooms or booths equipped with an appropriate exhaust system?

9. Is employee exposure to welding fumes controlled by ventilation, use of respirators, exposure time limits, or other means?

10. Are welders and nearby workers provided with flash shields during welding operations?

11. If forklifts and other vehicles are used in buildings or other enclosed areas, are the carbon monoxide levels kept below maximum acceptable concentration?

Noise Levels

12. Has it been determined that noise levels in the facilities are within acceptable levels?

13. Are steps being taken to use engineering controls to reduce excessive noise levels?

Hazardous Materials

14. Are caution labels and signs used to warn of hazardous substances (e.g., asbestos) and biohazards (such as blood-borne pathogens)?

15. Are wet methods used, when practicable, to prevent the emission of airborne asbestos fibers, silica dust, and similar hazardous materials?

16. Are proper precautions being taken when handling asbestos and other fibrous materials?

Operations

17. Is the possible presence of asbestos determined prior to the beginning of any repair, demolition, construction, or reconstruction work? ☐ ☐ ☐
18. Are asbestos-covered surfaces kept in good repair to prevent release of fibers? ☐ ☐ ☐
19. Are engineering controls examined and maintained or replaced on a scheduled basis? ☐ ☐ ☐
20. Is vacuuming with appropriate equipment used whenever possible rather than blowing or sweeping dust? ☐ ☐ ☐

Ventilation

21. Is the volume and velocity of air in each exhaust system sufficient to gather dusts, fumes, mists, vapors, or gases to be controlled, and convey them to a suitable disposal point? ☐ ☐ ☐
22. Are exhaust inlets, ducts, and plenums designed, constructed, and supported to prevent collapse or failure of any part of the system? ☐ ☐ ☐
23. Are clean-out ports or doors provided at intervals not to exceed 12 ft (3.6576 m) in all horizontal runs of exhaust ducts? ☐ ☐ ☐
24. Where two or more different operations are being controlled through the same exhaust system, could the combination of substances involved create a fire, explosion, or chemical reaction hazard in the duct? ☐ ☐ ☐
25. Is adequate makeup air provided to areas where exhaust systems are operating? ☐ ☐ ☐
26. Is the source point for makeup air located so that only clean, fresh air, free of contaminants will enter the work environment? ☐ ☐ ☐
27. Where two or more ventilation systems serve a work area, is their operation such that one will not offset the functions of the other? ☐ ☐ ☐
28. Are grinders, saws, and other machines that produce respirable dusts vented to an industrial collector or central exhaust system? ☐ ☐ ☐
29. Are all local exhaust ventilation systems designed to provide sufficient air flow and volume for the application, and are ducts not plugged? ☐ ☐ ☐
30. Are belts properly secured to prevent slipping? ☐ ☐ ☐

31. Is personal protective equipment (PPE) provided, used, and maintained wherever required? ☐ ☐ ☐

32. Are there written standard operating procedures for selecting and using respirators where needed? ☐ ☐ ☐

Sanitary Facilities

33. Are restrooms and washrooms kept clean and sanitary? ☐ ☐ ☐

34. Is all water provided for drinking, washing, and cooking potable? ☐ ☐ ☐

35. Are all outlets for water that is not suitable for drinking clearly identified? ☐ ☐ ☐

36. When non-potable water is piped through a facility, are outlets or taps posted to alert employees that water is unsafe and not to be used for drinking, washing, or personal use? ☐ ☐ ☐

37. Are employees' physical capacities assessed before they are assigned to jobs requiring heavy work? ☐ ☐ ☐

38. Are employees instructed in the proper manner for lifting heavy objects? ☐ ☐ ☐

39. Where heat is a problem, have all fixed work areas been provided with spot cooling or air conditioning? ☐ ☐ ☐

40. Are employees screened before assignment to areas of high heat to determine if their health might make them more susceptible to having an adverse reaction? ☐ ☐ ☐

41. Are employees, working on streets and roadways who are exposed to the hazards of traffic, required to wear high-visibility clothing such as bright-colored (traffic orange) warning vests? ☐ ☐ ☐

42. Are exhaust stacks and air intakes located so that nearby contaminated air will not be recirculated within a building or other enclosed area? ☐ ☐ ☐

43. Is equipment producing ultraviolet radiation properly shielded? ☐ ☐ ☐

44. Are universal precautions observed where occupational exposure to blood or other potentially infectious materials can occur and in all instances where differentiation of types of body fluids or potentially infectious materials is difficult or impossible? ☐ ☐ ☐

Operations

Piping Systems

45. When non-potable water is piped through a facility, are outlets or taps posted to alert employees that the water is unsafe and not to be used for drinking, washing, or other personal use? ☐ ☐ ☐

46. When hazardous substances are transported through above-ground piping, is each pipeline identified at points where confusion could introduce hazards to employees? ☐ ☐ ☐

47. When pipelines are identified by color-painted bands or tapes, are the bands or tapes located at reasonable intervals and at each outlet, valve, or connection, and are all visible parts of the line so identified? ☐ ☐ ☐

48. When pipelines are identified by color, is the color code posted at all locations where confusion could introduce hazards to employees? ☐ ☐ ☐

49. When the contents of pipelines are identified by name or name abbreviation, is the information readily visible on the pipe near each valve or outlet? ☐ ☐ ☐

50. When pipelines carrying hazardous substances are identified by tags, are the tags constructed of durable materials, the message printed clearly and permanently, and are tags installed at each valve or outlet? ☐ ☐ ☐

51. When pipelines are heated by electricity, steam, or other external source, are suitable warning signs or tags placed at unions, valves, or other serviceable parts of the system? ☐ ☐ ☐

Chapter 2

Responsibilities

Construction crane failures (as well as other construction accidents) are, by definition, preventable—they don't really represent accidents as much as negligence.

The failure of a crane to perform efficiently and its subsequent role in someone's death generally never lies with the equipment failure itself. Instead, it belongs to those who built, provided for or contracted for, and operate that equipment that has turned too frequently into a common killer.

Using cranes, derricks, and hoists in construction work does not have to be as dangerous as it has become.

A combination of factors have contributed to the dangers of rigging:

- Ineffective regulations governing construction cranes (which have wildly varied standards for operational safety)
- Contractor who is doing the job, or the company that leased and provided the crane

2.1 Safety Organization

A responsible safety program requires monitoring hoisting and rigging operations every step of the way to ensure that they are performed safely.

The project contractor or facility manager who initiates the contract for vendor-owned and -operated equipment is responsible for ensuring that the vendor equipment and personnel meet pertinent hoisting and rigging safety requirements of OSHA 29 CFR 1910 and 29 CFR 1926. And, the contractor or manager MUST designate a competent staff member to coordinate compliance with pertinent safety requirements.

Training

Riggers and signalmen are a critical part of crane operations, and they need to receive proper training to wok on the specific pieces of equipment they are responsible for maintaining. In addition, they MUST receive training similar to that provided to operators—as they are at risk while working with crane loads. Riggers should always use taglines to guide a lifted load into place as well as to help ensure that a lifted load will not swing into the boom, which can cause boom failure. Riggers also should be trained to identify hazardous conditions in crane rigging, such as wire rope deformation, strain, binding, or kinking.

It is imperative (and required by OSHA and ANSI) that crane maintenance personnel be properly trained to work on the specific pieces of equipment they are responsible for maintaining.

Maintenance

An aggressive routine of scheduled and preventive maintenance and repair MUST include all mechanical, electrical, and hydraulic systems, in addition to crane structures or areas exposed to high mechanical stresses. To ensure proper function and structural integrity of equipment, maintenance personnel MUST also understand how individual components interact with the entire crane, and what deleterious effects can result from their improper performance.

Inspections

OSHA requires inspection of machinery and other equipment (including wire ropes) by a competent person before each use, and rigging equipment should be inspected prior to use on each shift, and as necessary during its use to ensure that it is safe. Types of inspections include: initial, pre-use, frequent, periodic, and third-party inspections. Any defective rigging equipment MUST be immediately removed from service.

2.2 Crane-Specific Responsibilities

A number of different parties have a variety of responsibilities for safe crane operations.

Crane manufacturers

- Monitor the field performance of their cranes.
- Monitor maintenance/spare parts requests to assess whether any design aspect might be contributing to product failures or problems.

Crane rental companies

- Provide equipment that meets appropriate standards and is well maintained.
- Maintain records of each crane (serial number, parts manual, maintenance schedule, modifications, etc.) for the lifetime of the crane.

Construction contractors and subcontractors renting the equipment

- Ensure the crane is in proper operating condition.
- Have effective crane safety inspection procedures in place.
- Notify the crane rental company and manufacturer when problems are identified.

Note: All levels of contractor management in companies working with cranes MUST be committed to safe crane operations and MUST FOLLOW safe lifting guidelines at all times, despite timeline pressures, which may encourage companies to limit the use of safe procedures.

Project supervisors

- Ensure that employees under their supervision receive the required training and are certified and licensed to operate the cranes and hoists in their areas.
- Provide training for prospective crane and hoist operators.
- Evaluate crane and hoist trainees using a Crane Safety Checklist and submitting a Qualification Request Form to the Engineering Department/Division Head, Center director, or designee to obtain the operator's license.
- Ensure that hoisting equipment is inspected and tested monthly by a responsible, designated individual, and that rigging equipment is inspected annually.
- Ensure that inspection results are documented on-site.

Crane and hoist operators

- Restrict crane operations with which they have attained technical and performance proficiency.
- Operate hoisting equipment safely and NEVER operate crane or hoist in conditions that could compromise proper operation or mechanical integrity of the equipment.

- Conduct functional tests prior to using the equipment.
- Select and use rigging equipment appropriately.
- Have a valid operator's license on their person while operating cranes or hoists.
- Participate in a medical certification program (if required).

Safety engineer (staff or third party)
- Perform annual maintenance and inspection of all cranes and hoists that are not covered by an individual maintenance agreement.
- Conduct periodic and special load tests of cranes and hoists.
- Maintain written records of inspections and tests, and provide copies of all inspections and test results to project management.
- Inspect and load test cranes and hoists following modification or extensive repairs (such as, a replaced cable or hook, or structural modification).
- Schedule a nondestructive test and inspection for crane and hoist hooks at the time of the periodic load test.
- Schedule testing and inspection before using new replacement hooks and other hooks suspected of having been overloaded.

Note: Evaluation, inspection, and testing may include, but are not limited to visual, dye penetrant, and magnetic particle techniques referenced in ASME B30.10—Hooks, Inspection, and Testing.

- Maintain all manuals for cranes and hoists in a central file for on-site reference.
- Conduct training for all crane and hoist operators.
- Issue licenses to crane and hoist operators, for fixed location equipment.
- Maintain, on site, a current list of designated operators.
- Repairs (such as a replaced cable or hook, or structural modification).

Engineering safety office
- Verify monthly test and inspection reports—periodically.
- Interpret crane and hoist safety rules and standards.
- Provide management planning and technical assistance.

2.3 Project Responsibilities

Safe operating requirements

All workers who use any crane or hoist MUST have an operator's license. Workmen permitted to do rigging tasks—NOT just maintenance personnel, but also workers at the lowest level—should be tested and certified as competent on the use, inspection, and maintenance of rigging equipment.

- Certification—includes a theoretical and practical test with NO limit in weight—heavy weights and difficult loads to be classified as critical tasks with the relevant risks identified and safe operating practices drafted to address risks
- Theoretical test—based on an accredited training program
- Practical test—based on prescribed outcomes, relevant to either maintenance or operational personnel, as determined by the initial risk assessments
- Retesting of competent persons—to be done as and when required
- Hazard identification—by line supervisor regarding rigging activities and equipment
- Relevant training and certification—to be given to maintenance and operation personnel applicable to tasks to be performed

Project contractor/facility manager

- Establish programs based on rigging equipment manufacturers' specifications and limitations for the operation, maintenance, and inspection of equipment. Where manufacturers' specifications are not available, operation limitations and maintenance and inspection requirements assigned to the equipment MUST be based on determinations of a qualified person competent in this field and such determinations will be appropriately documented and disseminated to equipment operators and maintenance and test personnel.
- Appoint supervisors of rigging equipment maintenance and operation—only persons found to be competent with emphasis on responsibility for maintaining rigging equipment logbooks.
- Authorize (in writing) persons qualified to perform rigging tasks.

Hoisting and rigging project manager

- Ensure consistency in implementation and interpretation of the hoisting and rigging program across the worksite.

- Be the authority having jurisdiction over interpretation of the hoisting and rigging program.
- Help ensure that the pertinent hoisting and rigging issues are identified during subsequent investigations or critiques.
- Help ensure that identified hoisting and rigging issues are adequately addressed in corrective actions or lessons learned issued.
- Help ensure that any Occurrence Reporting and Processing System report or official lessons learned issued adequately addresses the hoisting and rigging aspects of and hoisting and rigging corrective actions and lessons learned for the event.
- Periodically assess line management implementation of the hoisting and rigging program at the worksite.
- Assist line organizations, when requested, in matters relating to hoisting and rigging, including H&R surveillances, reviews of critical or major lift procedures or work packages, hostile environment plans, and participation in the hoisting and rigging aspects of readiness assessments, and operational readiness reviews, etc.
- Assist site contractors, when requested, in addressing issues related to compliance with, implementation of, or interpretation of the hoisting and rigging program.
- Maintain the project site hoisting and rigging manual.
- Be the project authority for the review and approval of revisions to the hoisting and rigging program.
- Ensure the program and line organizations are kept up-to-date with the latest hoisting and rigging program changes, bulletins, or important issues applicable to their organizations.

Supervisor of hoisting and rigging operations

- Ensure that employees under their supervision receive the required training and are certified and licensed to operate the cranes and hoists in their areas.
- Assign ONLY qualified personnel to operate equipment and perform hoisting and rigging tasks.
- Provide training for prospective crane and hoist operators; conducted by a qualified, designated instructor who is a licensed crane and hoist operator and a full-time company employee.
- Evaluate crane and hoist trainees using the Crane Safety Checklist and submitting the Qualification Request Form to the Safety Office to obtain the operator's license.

- Ensure that hoisting equipment is inspected and tested monthly by a responsible individual; and that rigging equipment is inspected annually.
- Ensure equipment is operated safely.
- Use planned and approved hoisting and rigging instructions when necessary—always for critical lifts.
- Properly tag equipment found to be unsafe or requiring restrictive use.
- Notify equipment custodian of equipment problems.
- Assign a designated leader to hoisting and rigging operations that require more than one person.

Supervisor of inspection, maintenance, and repair

- Inspect, maintain, and repair hoisting and lifting equipment and rigging components, including:
  - Ensure hoisting and lifting equipment and rigging components are inspected, maintained, and repaired by qualified personnel.
  - Ensure inspectors, maintenance, and repair personnel have the tools to safely accomplish their work.
  - Ensure responsible inspectors and maintenance and test personnel have access to adequate information, as applicable, including operating instructions; maintenance, repair, and parts information furnished by the manufacturer or the responsible maintenance/engineering organization; manufacturer's recommendations as to points and frequency of lubrication, maintenance of lubrication levels, and types of lubricant to be used; and maintenance or repair procedures from the manufacturer or responsible maintenance/engineering organization.
- Ensure document inspection, maintenance, and repair activities in accordance with project requirements.
- Ensure personnel responsible for inspection or maintenance are familiar with the applicable contents of the manual furnished with equipment.

Designated leader for critical lifts

Management MUST assign a designated leader for critical lifts—a crew member or any qualified person to ensure that

- A critical lift procedure is prepared.
- The critical lift procedure is properly approved before implementing.

- A documented pre-lift meeting is held and personnel understand how the job will be done.
- Management provides qualified personnel (such as operators, riggers, flagman, and designated leader).
- Proper equipment and hardware are identified in the critical lift procedure.
- The lifting operation is directed by a designated leader to ensure that the job is done safely and efficiently.
- Involved personnel are familiar with, and follow, the critical lift procedure.

After the critical lift is completed, critical lift documentation MUST be transmitted to the project manager for whom the lift was done with the designated leader advising responsible personnel that this documentation is subject to audit for 1 year.

Designated leader for hoisting and rigging

Contractor shall appoint a designated leader for hoisting and rigging activities that involve more than one person. (Normal forklift truck material handling operations are not considered hoisting and rigging activities and do not require a designated leader.)

The designated leader may be the operator, a crew member, or any qualified person with responsibility to:

- Ensure that a flagman or signaler, if required, is assigned and identified to the hoist/crane/forklift operator.

Note: Rigging (e.g., wire rope, synthetic, chain, shackles) may be components supplied with front-end attachments. When integrated into front-end attachments, such components do not invoke hoisting and rigging designated leader appointment.

- Ensure that management provides qualified personnel and they understand how the job is to be done.
- Ensure that the weight of the load is determined, that the proper equipment and hardware are selected and inspected, and that the capacity of the lifting device is not exceeded.
- Ensure that the equipment is properly set up and positioned.
- Examine the work area for hazardous or unsafe conditions.
- Direct the lifting operation to ensure that the job is done safely and efficiently.

- Ensure that the job is stopped when any potentially unsafe condition is recognized.
- Ensure prompt reporting of any injury or accident emergency that occurs.
- Take charge of the accident scene pending arrival of emergency services personnel.

Equipment operator

Operators perform the following activities:

- Have a valid operator's license on their person while operating cranes or hoists.
- Operate hoisting equipment safely.
- Ensure equipment operating guidelines and, for mobile cranes, the load charts.
- Ensure preuse and frequent equipment inspection.
- Ensure inspections are current from an inspection sticker, other documentation, or verbal confirmation from the equipment custodian.
- Conduct functional tests prior to using the equipment.
- Select and use rigging equipment appropriately.
- Ensure that the load will not exceed the rated capacity of the equipment.
- Abide by any restrictions placed on the use of the equipment.
- Participate in the company's medical certification program, as required.

Rigger

Riggers perform the following activities:

- Ensure that rigging equipment and materials have the required capacity for the job.
- Ensure that all items are in good condition, currently qualified (up-to-date inspection), and are properly used.
- Verify that rigging equipment and material are in compliance with the procedure (if applicable).
- Confirm that the load path is clear of personnel and obstacles.

Equipment custodian

Management shall designate an individual having custodial responsibility for each crane, hoist, lift truck, or other hoisting and rigging equipment that require scheduled maintenance, inspection, and record keeping to perform the following activities:

- Verify that operating equipment is properly maintained and maintenance, inspection, and testing of the equipment remain current.
- Ensure that records of the maintenance, repair, inspection, and testing are available for audit in a maintenance file.
- Verify that equipment is properly tagged and, if necessary, removed from service when discrepancies are found during inspection or operation.

Note: It is important that equipment users know how to contact the equipment custodian. Devise a method so that equipment users can easily identify and contact the equipment custodian. More specific, and possibly additional, responsibilities may be stated depending on specific project circumstances.

Maintenance operations department

- Maintain all manuals for cranes and hoists in a central file for reference.
- Perform annual maintenance and inspection of all company cranes and hoists that are not covered by a program with maintenance responsibility.
- Conduct periodic and special load tests of cranes and hoists.
- Maintain written records of inspections and tests, and provide copies of all inspections and test results to facility managers and building coordinators who have cranes and hoists on file.
- Inspect and load test cranes and hoists following modification or extensive repairs (such as a replaced cable or hook, or structural modification).
- Schedule a nondestructive test and inspection for crane and hoist hooks at the time of the periodic load test.
- Test and inspect before using new replacement hooks and other hooks suspected of having been overloaded.

Note: Evaluation, inspection, and testing may include, but is not limited to visual, dye penetrant, and magnetic particle techniques referenced in ASME B30.10—Hooks, Inspection, and Testing.

Sources:
Construction Safety Association, Ontario, Canada (www.csao.org)
Stanford University, Menlo Park, CA (www.slac.stanford.edu)
Texas A&M University, College Station, TX (www.tamu.edu)
U.S. Department of Energy, Pacific Northwest Laboratory, Richland, WA (www.pnl.gov)

Chapter 3

Procedures

Before conducting an ordinary lift and operating material handling equipment to complete work, qualified rigging engineers emphasize preparing a lift assessment. All ordinary lifts requiring the use of material handling equipment should follow work-planning procedures as recommended by OSHA to ensure safe lifting and operating practices.

3.1 Planning and Preparation

Brookhaven National Laboratory engineers have developed the following set of procedure guidelines applicable for ordinary hoisting and lifting procedures.

- Determine the type of lift by conducting a lift assessment.
  - Evaluate lift.
  - Ensure training is complete.
  - Ensure material handling equipment is certified.
  - Present plan for review.
  - Start activity when work package is complete and approved.
  - Keep nonparticipants out of work control zone.
  - Approve work planning.
  - Hold pre-lift meeting.
  - Conduct lift.
- Certify material handling equipment for use.
  - Inspect and certify (load test) new equipment not certified by manufacturer.
  - Review and inspect lifting device.
  - Ensure equipment is used only for its intended purpose.

Operations

- Inspect and maintain lifting and material handling equipment including:
  - Inspection and proof test requirements of rigging and material handling hardware
  - Inspection and proof test requirements of rigging hooks
  - Inspection, proof test requirements, and operating practices for slings
  - Inspection and proof test requirements of below-the-hook lifting devices
  - Inspection, load test requirements, and operating practices for overhead cranes and hoists
  - Inspection, load test requirements, and operating practices for mobile cranes and boom trucks
  - Maintenance of lifting and material handling equipment

Source: Brookhaven National Laboratory, Brookhaven, NY (www.bnl.gov)

Regulations

This checklist is designed to assist you in conducting self-inspections for hoisting and lifting operations. By no means, however, is the list all-inclusive. Add or delete those portions or items that do not apply to particular operations, but MAKE SURE to carefully consider each item before changing anything.

Refer to OSHA standards for complete and specific standards that may apply to specific work situations.

HOISTING AND RIGGING 29 CFR 1926.753	YES	NO	N/A
1. Are hoisting and rigging operations preplanned to reduce employee exposures to overhead loads? [29 CFR 1926.752(d)]	☐	☐	☐
2. Does a competent person perform a pre-shift inspection? [29 CFR 1926.753(c)]	☐	☐	☐
3. Does a qualified rigger (rigger who is also a qualified person) inspect the rigging prior to each shift? [29 CFR 1926.753(c)(2)]	☐	☐	☐

Crane Operations

4. Are operators PROHIBITED to use cranes to hoist personnel unless all provisions of 29 CFR 1926.550 are met except
   - When the erection, use, and dismantling of conventional means of reaching the worksite, such as a personnel hoist, ladder, stairway, aerial lift, elevating work platform, or scaffold, would be more hazardous or is not possible because of structural design or worksite conditions? [29 CFR 1926.550(g)(2)] ☐ ☐ ☐

5. Are the requirements of 29 CFR 1926.753(d) followed when riggers work under loads (allowed in specified instances), such as engaged in the initial connection of the steel or the hooking or unhooking of the load? [29 CFR 1926.753(d)(1)(ii)] ☐ ☐ ☐

6. Are workers PROHIBITED from deactivating hooks unless a qualified rigger determines it is safer to place steel without them, or equivalent protection provided in a site-specific erection plan? [29 CFR 1926.753(c)(5)] ☐ ☐ ☐

Note: The load must be rigged by a qualified rigger [29 1956.753(d)].

7. Do operators, responsible for operations under their control, use their authority to stop and refuse to handle loads until safety has been assured? [29 CFR 1926.753(c)(2)(iv)] ☐ ☐ ☐

Falling Object Protection 29 CFR 1926.759

8. Are all materials, equipment, and tools, NOT being used, secured against accidental displacement? [29 CFR 1926.759(a)] ☐ ☐ ☐

9. Does the controlling contractor bar other construction processes below steel erection unless overhead protection is provided for the employees working below? [20 CFR 1926.759(b)] ☐ ☐ ☐

Operations

Fall Protection   29 CFR 1926.760

10. Are all employees protected at heights over 15 ft (4.57 m) and up to 30 ft (9.14 m) with a personal fall arrest system, positioning device system, or fall restraint system?
    - Do they wear the equipment necessary to be able to be tied off, or are they provided with other means of protection from fall hazards?
    [29 CFR 1926.760(b)(3)]
11. Is a controlled decking zone established in that area of the structure over 15 ft (4.57 m) and up to 30 ft (9.14 m) above a lower level where metal decking is initially being installed and forms the leading edge of a work area?
    [29 CFR 1926.760(c)]
12. Are boundaries of a controlled decking zone designated and clearly marked?
    [29 CFR 1926.760(c)(3)]

Note: The controlled decking zone MUST NOT be more than 90 ft (27.4 m) wide and deep from any leading edge, and MUST be marked by the use of control lines or the equivalent.

13. Does the controlling contractor accept the responsibility for choosing fall protection equipment used by riggers?
    [20 CFR 1926.760(e)]

Training

14. Is lifting and hoisting training of riggers conducted by qualified person(s)?
    [29 CFR 1926.761(a)]
15. Are special training programs conducted for multiple lift rigging, connectors, and controlled decking zones?
    [29 CFR 1926.761(c)(1) through (3)(ii)]

SITE-SPECIFIC ERECTION PLAN   29 CFR 1926.752   YES   NO   N/A

1. Has a qualified person developed and implemented a site-specific erection plan when the controlling contractor has decided to use alternative means and methods to protect employees from three specific hazards?

- Is that qualified person available at the worksite? ☐ ☐ ☐
  [29 CFR 1926.752(e)]

Visual Site Inspection

2. Did the controlling contractor provide adequate road ☐ ☐ ☐
   access on the site for the delivery and movement of
   derricks, cranes, trucks, steel erection materials, and
   other equipment?
   - Did the controlling contractor provide means and ☐ ☐ ☐
     methods for pedestrian and vehicular control?
     [29 CFR 1926.752(c)(1)]

Note: This requirement DOES NOT apply to roads outside of the construction site.

3. Did the controlling contractor provide a firm, properly ☐ ☐ ☐
   graded, drained area, readily accessible to the work
   with adequate space for the safe storage of materials
   and safe operation of the erectors' equipment?
   [29 CFR 1926.752(c)(2)]

4. Did the controlling contractor either bar other ☐ ☐ ☐
   construction processes below steel erection or provide
   overhead protection for the employees below?
   [29 CFR 1926.759(b)]

Note: This relates ONLY to protection from falling objects other than materials being hoisted.

3.2 Inspections

Crane operators MUST perform pre-use inspection (once every shift) lifting or hoisting machinery scheduled for use. Any equipment found to be unsatisfactory MUST be removed from service immediately.

Operators MUST notify the equipment custodian, who will submit a service request to the crane operator. Custodians MUST keep completed checklists on file for a minimum of 5 years.

This checklist is designed to assist the rigging contractor in conducting self-inspections for hoisting and lifting operations. By no means, however, is the list all-inclusive. Add or delete those portions or items that do not apply to particular operations, but MAKE SURE to carefully consider each item before changing anything.

Refer to OSHA standards for complete and specific standards that may apply to specific work situations.

	YES	NO	N/A
1. Are unit facilities annually inspected by the Building Safety Department and copies of discrepancy reports received by the unit returned to management with corrective actions noted as required?	☐	☐	☐
2. Have personnel elected or appointed to serve on unit safety committees, i.e., unit managers and supervisors, attended a facility inspections course presented by a qualified person?	☐	☐	☐
3. Are in-house facility safety inspections being performed and documented?	☐	☐	☐
4. Are all fire extinguishers serviced annually by a qualified technician and appropriately tagged?	☐	☐	☐
5. Are all fire extinguishers inspected monthly and tags attached to each unit dated and initialed to document inspections?	☐	☐	☐

Lift and Inspection

	YES	NO	N/A
6. Does the responsible manager or designee determine the type of lift by conducting a lift assessment:			
• Ordinary lift?	☐	☐	☐
• Pre-engineered lift?	☐	☐	☐
• Critical lift?	☐	☐	☐
7. Does the operator conduct a pre-use inspection of the equipment to be used and ensure that the equipment is approved for use in the particular work area?	☐	☐	☐
8. Does the responsible manager or designee (e.g., professional engineer, person-in-charge) evaluate the proposed lift or material handling requirement in accordance with work planning requirements and authorize the activity or lift?	☐	☐	☐
9. Does the responsible manager or designee ensure all personnel performing mechanical material handling, hoisting, and rigging activities, operating powered industrial trucks (forklifts), or other lifting equipment have completed the required training (i.e., Basic Rigging Course, Forklift Operator Course)?	☐	☐	☐

10. Does the responsible manager or designee ensure the equipment to be used is certified for the use in that area by inspecting the tag on the equipment for a current inspection date? ☐ ☐ ☐

Note: If the inspection due date has passed, DO NOT use the equipment and remove it from service until reinspected and certified for use.

11. Does the responsible manager or designee (or the contractor, if applicable) present the plan to the supervisor or cognizant person-in-charge for review with the person(s) who will perform the task(s) using the material handling equipment, noting the precautions required to safely complete the task(s)? ☐ ☐ ☐

Note: All precautions required to safely complete the task(s) MUST be noted in the pre-lift meeting, and as a minimum, cover the following:

Intended steps in the lift sequence
Work control zone, load path, and unloading area
Potential hazards, hazardous situations in the handling area
How the hazards will be mitigated
Stop Work Authority
Load securement and transportation issues.

The responsible manager or designee (or the contractor, if applicable) should allow an adequate amount of time in the schedule to permit load testing and feedback to be addressed as needed.

12. Does the activity begin once the work package is complete and accepted, the material handling equipment meets inspection requirements, and qualified operators are determined and identified? ☐ ☐ ☐

Note: In certain situations, it may be prudent to have experienced personnel present when the activity begins. If the intended work is to be done by "qualified personnel," the responsible line organization should make that part of the plan and verify that such personnel are performing the work.

13. Does the work crew keep nonparticipants out of the work control zone (danger zone), whether indoors or outdoors—from the onset of the work, whether it is a hoisting and rigging operation, a forklift operation, or other operation? ☐ ☐ ☐

Note: Designated personnel in charge of forklift, mobile crane, overhead crane, and aerial lifts MUST ensure that work control zones are set up.

14. Does line management approve work planning and related activities? ☐ ☐ ☐
15. Do contractors obtain approval from the project or facility engineering, hoisting, and rigging inspector? ☐ ☐ ☐

Note: Records MUST be maintained by line management.

16. Is the lift conducted as planned, and is the load properly secured? ☐ ☐ ☐
17. Do managers ensure that personnel when operating material handling equipment are wearing hard hats at all times? ☐ ☐ ☐
18. Are there are any problems during a lift that cause a departure from the lifting plan? ☐ ☐ ☐

Note: If there are problems, abort the lift, redo the planning and review, and conduct the lift at another time.

GENERAL WORK ENVIRONMENT   YES   NO   N/A

1. Are all worksites clean, sanitary, and orderly? ☐ ☐ ☐
2. Are work surfaces kept dry or appropriate means taken to ensure the surfaces are slip-resistant? ☐ ☐ ☐
3. Are all spilled hazardous materials or liquids, including potentially infectious materials, cleaned up immediately and according to proper procedures? ☐ ☐ ☐
4. Is combustible scrap, debris, and waste stored safely and removed from the worksite promptly? ☐ ☐ ☐
5. Is all regulated waste, as defined in the OSHA blood-borne pathogens standard (29 CFR 1910.1030) discarded according to federal, state, and local regulations? ☐ ☐ ☐
6. Are accumulations of combustible dust routinely removed from elevated surfaces, including the overhead structure of buildings, etc.? ☐ ☐ ☐
7. Is combustible dust cleaned up with a vacuum system to prevent the dust going into suspension? ☐ ☐ ☐

8. Is metallic or conductive dust prevented from entering or accumulating on or around electrical enclosures or equipment? ☐ ☐ ☐

9. Are covered metal waste cans used for oily and paint-soaked waste? ☐ ☐ ☐

10. Are all oil and gas fire devices equipped with flame failure controls that will prevent flow of fuel if pilots or main burners are not working? ☐ ☐ ☐

11. Are paint spray booths, dip tanks, etc. cleaned regularly? ☐ ☐ ☐

12. Are the minimum number of toilets and washing facilities provided? ☐ ☐ ☐

13. Are all toilets and washing facilities clean and sanitary? ☐ ☐ ☐

14. Are all work areas adequately illuminated? ☐ ☐ ☐

15. Are pits and floor openings covered or otherwise guarded? ☐ ☐ ☐

EMPLOYER POSTING   YES   NO   N/A

1. Is the required Department of Labor (DOL) workplace poster displayed in a prominent location where all employees are likely to see it? ☐ ☐ ☐

2. Are emergency telephone numbers posted where they can be readily found in case of emergency? ☐ ☐ ☐

3. Where employees may be exposed to any toxic substances or harmful physical agents, is appropriate information concerning employees made available, especially:
   - Medical and exposure records? ☐ ☐ ☐
   - Material Safety Data Sheets? ☐ ☐ ☐

4. Are signs concerning "Exiting from buildings," room capacities, floor loading, biohazards, exposures to x-ray, microwave, or other harmful radiation or substances posted where appropriate? ☐ ☐ ☐

5. Is the Summary of Occupational Illnesses and Injuries posted in the month of February? ☐ ☐ ☐

Record Keeping

6. Are all occupational injury or illnesses, except minor injuries, requiring only first aid being recorded as required on the log? ☐ ☐ ☐

7. Are employee medical records and records of employee exposure to hazardous substances or harmful physical agents up-to-date and in compliance with current LOCAL standards? ☐ ☐ ☐

8. Are employee training records kept and accessible for review by employees, when required by DOL standards? ☐ ☐ ☐

9. Have arrangements been made to maintain required records for the legal period of time for each specific type record? ☐ ☐ ☐

Note: Some records must be maintained for at least 40 years.

10. Are operating permits and records up-to-date for such items as elevators, air pressure tanks, liquefied petroleum gas tanks, etc.? ☐ ☐ ☐

SAFETY AND HEALTH PROGRAM   YES   NO   N/A

1. Do you have an active safety and health program in operation that deals with general safety and health program elements as well as the management of hazards specific to the particular worksite? ☐ ☐ ☐

2. Is one person clearly responsible for the overall activities of safety and health program? ☐ ☐ ☐

3. Is there a safety committee or group made up of management and labor representatives that meets regularly and reports in writing on its activities? ☐ ☐ ☐

4. Do you have a working procedure for handling in-house employee complaints regarding safety and health? ☐ ☐ ☐

5. Are employees kept advised of the successful effort and accomplishments that the safety committee has made in ensuring that they will have a workplace that is safe and healthful? ☐ ☐ ☐

MEDICAL SERVICES AND FIRST AID

	YES	NO	N/A
1. Is there a hospital, clinic, or infirmary for medical care in proximity to workplace?	☐	☐	☐
2. If medical and first-aid facilities are not in proximity to your workplace, is at least one employee on each shift currently qualified to render first aid?	☐	☐	☐
3. Have all employees who are expected to respond to medical emergencies as part of their work	☐	☐	☐
▪ Received first-aid training?	☐	☐	☐
▪ Had hepatitis B vaccination made available to them?	☐	☐	☐
▪ Had appropriate training on procedures to protect them from blood-borne pathogens, including universal precautions?	☐	☐	☐
▪ Had appropriate personal protective equipment available and understood how to use it to protect against exposure to blood-borne diseases?	☐	☐	☐
4. Where employees have had an exposure incident involving blood-borne pathogens, did you provide an immediate postexposure medical evaluation and follow-up?	☐	☐	☐
5. Are medical personnel readily available for advice and consultation on matters of employees' health?	☐	☐	☐
6. Are emergency phone numbers posted?	☐	☐	☐
7. Are first-aid kits easily accessible to each work area, with necessary supplies available, periodically inspected, and replenished as needed?	☐	☐	☐
8. Have first-aid kit supplies been approved by a physician, indicating that they are adequate for a particular area or operation?	☐	☐	☐
9. Are means provided for quick drenching or flushing of the eyes and body in areas where corrosive liquids or materials are handled?	☐	☐	☐

Note: Pursuant to an OSHA memorandum of July 1, 1992, employees who render first aid only as a collateral duty DO NOT have to be offered pre-exposure hepatitis B vaccine, only if the employer puts the following requirements into the company's exposure control plan and implements them:

- Record all first-aid incidents involving the presence of blood or other potentially infectious materials before the end of the work shift during which the first-aid incident occurred.
- Comply with postexposure evaluation, prophylaxis, and follow-up requirements of the standard with respect to "exposure incidents," as defined by the standard.
- Train designated first-aid providers in the reporting procedure.
- Offer to initiate the hepatitis B vaccination series within 24 hours to all unvaccinated first-aid providers who have rendered assistance in any situation involving the presence of blood or other potentially infectious materials.

3.3 Standards (Consensus)

ANSI/ASME B30 Standards applicable to various cranes, hoists, lifting systems, and derricks covering construction, installation, operation, inspection, and maintenance of rigging equipment and systems.

B30.3-004 Construction Tower Cranes
Applies to construction tower cranes, powered by electric motors or internal combustion engines, and any variations thereof which retain the same fundamental characteristics.

B30.4-003 Portal, Tower, and Pedestal Cranes
Applies to the electric motor or internal-combustion engine-powered portal tower, and pedestal cranes, that adjust operating radius by means of a boom luffing mechanism or a trolley traversing a horizontal boom, that may be mounted on a fixed or traveling base, and to any variations thereof that retain the same fundamental characteristics. This standard applies only to portal tower and pedestal cranes utilizing a drum and wire rope for hoisting and which are used for hoisting work.

B30.5-2007 Mobile and Locomotive Cranes
Applies to crawler cranes, locomotive cranes, wheel-mounted cranes, and any variations thereof that retain the same fundamental characteristics. The scope includes only cranes of the above types that are basically powered by internal combustion engines or electric motors.

B30.9-2006   Slings
Applies to slings for lifting purposes, made from alloy steel chain, sewn synthetic webbing, wire rope, metal mesh, and synthetic fiber rope used in conjunction.

B30.10-2009   Hooks
Applies to hooks used in conjunction with equipment addressed in other volumes of the B30 Standard.

B30.11-2004   Monorails and Underhung Cranes
Applies to underhung cranes and monorail systems where load-carrying members, such as end trucks or carriers (trolleys), travel either on the external or internal lower flange of a runaway track section, single monorail track, crane bridge girder, or jib boom, including all curves, switches, transfer devices, lift and drop sections, and associated equipment. This volume includes provisions of both power-driven and hand-operated equipment on which the carriers are independently controlled.

B30.16-2007   Overhead Hoists (Underhung)
Applies to hand chain-operated chain hoists and electric- and air-powered chain and wire ropes hoists used for, but not limited to, vertical lifting and lowering of freely suspended, unguided, loads which consist of equipment and materials.

B30.20-2006   Below-the-Hook Lifting Devices
Applies to below-the-hook lifting devices, other than slings, used for attaching loads to hoist. Covers structural and mechanical lifting devices, vacuum lifting devices, operated close proximity lifting magnets, and remotely operated lifting magnets.

B30.22-2005   Articulating Boom Cranes
Applies to cranes with knuckle booms articulated by hydraulic cylinders, which are powered by internal combustion engines or electric motors and are mounted on a mobile chassis or stationary installation. Frequently, load hoist mechanism equipped machines are covered by this standard.

B30.23-2005   Personnel Lifting Systems
Applies to hoisting equipment and accessory equipment that is used to lift, lower, or transport personnel in a platform, by wire rope or chain, from hoist equipment, or by a platform that is mounted on a boom of the hoist equipment. The lifting of personnel is not allowed using some ASME 30 Standard equipment.

B30.26-2004 Rigging Hardware
Applies to detachable rigging hardware used for lifting purposes covering shackles, links, rings, swivels, turnbuckles, eyebolts, hoist rings, wire rope clips, wedge sockets, and rigging blocks.

Part 2

Inspections

Chapter 4

Lifting Equipment

4.1 Cranes and Derricks

Maintenance and safety inspections on lifting and hoisting equipment are an integral part of a rigging contractor's risk assessment program—as well as necessary to meet the legal requirements for periodic inspections. Equipment MUST be inspected by a competent person before and during use—with any deficiencies corrected before further use.

OPERATIONAL PROCEDURES

	YES	NO	N/A
1. Are only trained and authorized employees allowed to operate lifting equipment (such as cranes, derricks, hoists, forklifts, vehicle-mounted work platforms, and similar equipment)?	☐	☐	☐
2. Have employees who may operate aerial ladders or any other vehicle-mounted, elevating/rotating platforms that can be telescoped or articulated (or both) completed a Vehicle-Mounted Work Platforms Course?	☐	☐	☐
3. Do employees who operate equipment requiring a Commercial Driver's License (CDL) have a current license with the proper endorsements in their possession as required?	☐	☐	☐
4. Are all employees, who are required to hold a CDL, enrolled in a Random Drug Testing Program?	☐	☐	☐

Inspections

5. Have all employees who may have to use vehicle-mounted/portable cranes or hoists and other fixed, overhead lifting appliances completed a Crane and Hoist Safety Training Course conducted by a qualified third-party service provider? ☐ ☐ ☐

6. Do crane operators understand and use the load chart? ☐ ☐ ☐

7. Are operators able to determine the angle and length of the crane boom at all times? ☐ ☐ ☐

8. Are operators PROHIBITED to exceed rated loads—85% on outriggers or 75% on tires of the tipping load as determined by SAE Crane Stability Test Code J765a? ☐ ☐ ☐

Note: Rated loads are based on freely suspended loads. No attempt shall be made to drag a load horizontally on the ground in any direction.

9. Are crane operation ratings reduced to allow for adverse job conditions, such as soft or uneven ground, out-of-level conditions, high winds, side loads, pendulum action, jerking or sudden stopping of loads, hazardous conditions, experience of personnel, two machine lifts, traveling with loads, electric wires, etc.? ☐ ☐ ☐

Note: Side pull on boom or jib is HAZARDOUS. Derating of the crane's lifting capacity is required when wind speed exceeds 20 mi/h. The center of the lifted load must never be allowed to move more than 3 ft (91.44 cm) off the center line of the base boom section due to the effects of wind, inertia, or any combination of the two.

"Use 2 ft (61.91 cm) off the center line of the base boom for a two-section boom, 3 ft (91.44 cm) for a three-section boom, or 4 ft (1.22 m) for a four-section boom."

Note: The maximum load that can be telescoped is not definable because of variations in loadings and crane maintenance, but it is permissible to attempt retraction and extension if load ratings are not exceeded. Load ratings are dependent upon the crane being maintained according to manufacturer's specifications.

10. Are operators PROHIBITED to tip the crane to determine allowable loads? ☐ ☐ ☐

Note: When either radius or boom length, or both, are between listed values, ALWAYS use the smaller of the two listed load ratings.

Lifting Equipment   51

11. Are operators PROHIBITED to operate at longer radii than those listed on the applicable load-rating chart?   ☐ ☐ ☐

Note: The boom angles shown on the Capacity Chart give an approximation of the operating radius for a specified boom length. The boom angle, before loading, should be greater to account for boom deflection. It may be necessary to retract the boom if maximum boom angle is insufficient to maintain rated radius.

12. Are power telescoping boom sections ALWAYS extended equally?   ☐ ☐ ☐

Note: Rated loads include the weight of hook block, slings, and auxiliary lifting devices. Their weights shall be subtracted from the listed rated load to obtain the net load that can be lifted. When lifting over the jib, the weight of any hook block, slings, and auxiliary lifting devices at the boom head must be added to the load. When jibs are erected but unused, add two times the weight of any hook block, slings, and auxiliary lifting devices at the jib head to the load.

13. Are load handling devices, including hooks and hook blocks, kept away from boom head at all times?   ☐ ☐ ☐

14. Are truck crane operators PROHIBITED to lift with outrigger beams positioned between the fully extended and intermediate (pinned) positions?   ☐ ☐ ☐

15. Are operators PROHIBITED to use truck cranes not equipped with equalizing (bogie) beams between the rear axles for lifting "on tires"?   ☐ ☐ ☐

Note: Operators using truck cranes equipped with equalizing beams and rear air suspension should "dump" the air before lifting "on tires."

16. Are cranes and derricks restricted from operating within 10 ft (3.05 m) of any electrical power line?   ☐ ☐ ☐

17. Are rated load capacities, operating speed, and instruction posters visible to the operator?   ☐ ☐ ☐

18. Is the crane machinery and other rigging equipment inspected daily prior to use to make sure that it is in good condition?   ☐ ☐ ☐

19. Are accessible areas within the swing radius barricaded?   ☐ ☐ ☐

20. Are tag lines being used to prevent dangerous swing or spin of materials when raised or lowered by crane or derrick? ☐ ☐ ☐
21. Is a fire extinguisher of at least 5-BC rating provided on the crane? ☐ ☐ ☐
22. Are illustrations of hand signals to crane and derrick operators posted on the job site? ☐ ☐ ☐
23. Does the hook man use correct signals for the crane operator to follow? ☐ ☐ ☐
24. Are crane outriggers used as required?
25. Do crane platforms and walkways have anti-skid surfaces? ☐ ☐ ☐
26. Is broken, worn, or damaged wire rope removed from service? ☐ ☐ ☐
27. Are exhaust pipes guarded or insulated where employees might contact them? ☐ ☐ ☐
28. Are guardrails, handholds, and steps provided for safe and easy access to all areas of the crane? ☐ ☐ ☐
29. Are trolley and two-block limits on hammerhead tower cranes working? ☐ ☐ ☐
30. Have tower bolts been properly torqued? ☐ ☐ ☐
31. Have overload limits been tested and correctly set? ☐ ☐ ☐
32. Do personnel platforms suspended from crane hooks conform to OSHA requirements? ☐ ☐ ☐
33. Does the crane operation comply with manufacturer's specification? ☐ ☐ ☐

Source: Occupational Safety & Health Administration.

SAFE OPERATIONAL PROCEDURES	YES	NO	N/A
1. Are operators PROHIBITED to exceed Crane Load Ratings (CLR)?	☐	☐	☐
2. Are rated load capacities, operating speed, and instruction posters visible to the operator?	☐	☐	☐
3. Does the operator understand and use the load chart?	☐	☐	☐
4. Is the operator able to determine the angle and length of the crane boom at all times?	☐	☐	☐

5. Is the crane machinery and other rigging equipment inspected daily prior to use to make sure that it is in good condition? ☐ ☐ ☐

6. Are accessible areas within the swing radius barricaded? ☐ ☐ ☐

7. Are tag lines being used to prevent dangerous swing or spin of materials when raised or lowered by crane or derrick? ☐ ☐ ☐

8. Is a fire extinguisher of at least 5-BC rating provided on the crane? ☐ ☐ ☐

9. Are illustrations of hand signals to crane and derrick operators posted on the job site? ☐ ☐ ☐

10. Does the hook man use correct signals for the crane operator to follow? ☐ ☐ ☐

11. Are crane outriggers used as required? ☐ ☐ ☐

12. Do crane platforms and walkways have anti-skid surfaces? ☐ ☐ ☐

13. Is broken, worn, or damaged wire rope removed from service? ☐ ☐ ☐

14. Are exhaust pipes guarded or insulated where employees might contact them? ☐ ☐ ☐

15. Are guardrails, handholds, and steps provided for safe and easy access to all areas of the crane? ☐ ☐ ☐

16. Are trolley and two-block limits on hammerhead tower cranes working? ☐ ☐ ☐

17. Have tower bolts been properly torqued? ☐ ☐ ☐

18. Have overload limits been tested and correctly set? ☐ ☐ ☐

19. Do personnel platforms suspended from crane hooks conform to OSHA requirements? ☐ ☐ ☐

20. Does the crane operation comply with manufacturer's specification? ☐ ☐ ☐

Source: Occupational Safety & Health Administration.

CRAWLER, TRUCK, WHEEL, RINGER-MOUNTED CRANES YES NO N/A

1. Are outriggers fully extended and down (unless the manufacturer has specified an on-rubber rating)? ☐ ☐ ☐

Inspections

2. Are lattice boom cranes equipped with a boom angle indicator, load-indicating device, or a load moment indicator? ☐ ☐ ☐

3. Are lattice boom and hydraulic cranes equipped with a means for the operator to visually determine levelness? ☐ ☐ ☐

4. Are lattice boom and hydraulic cranes, except articulating booms cranes, equipped with drum rotation indicators located for use for the operator? ☐ ☐ ☐

5. Are lattice boom and hydraulic mobile cranes equipped with a boom angle or radius indicator within the operator's view? ☐ ☐ ☐

6. Are lattice boom cranes, with exception of duty cycle cranes, equipped with an anti-two-block device? ☐ ☐ ☐

7. When duty cycle machines are required to make a non-duty lift, is the crane equipped with an international orange warning device and is a signal person present? ☐ ☐ ☐

8. Are the following with the crane at all times:
   - Manufacturer's operating manual? ☐ ☐ ☐
   - Load-rating chart? ☐ ☐ ☐
   - Crane's logbook documenting use, maintenance, inspections, and tests? ☐ ☐ ☐
   - Operating manual for crane operator aids used on the crane? ☐ ☐ ☐

9. Are the following prepared on the project site:
   - Completed periodic inspection report prior to initial work? ☐ ☐ ☐
   - Preoperational checklist used for daily inspection? ☐ ☐ ☐
   - Written reports of the operational performance test? ☐ ☐ ☐
   - Written reports of the load performance test? ☐ ☐ ☐

10. Are all operators physically qualified to perform work? ☐ ☐ ☐

11. Are all operators qualified by written/oral and practical exam or by appropriate licensing agency for the type of crane they are to operate? ☐ ☐ ☐

12. Is the crane designed and constructed in accordance with industry standards? ☐ ☐ ☐

13. Is a hazard analysis for setup and set-down available? ☐ ☐ ☐
14. Are accessible areas within the swing radius of the rear of the crane barricaded? ☐ ☐ ☐
15. Are there at least three wraps of cable on the drum? ☐ ☐ ☐
16. Are the hoisting ropes installed in accordance with the manufacturer's recommendations? ☐ ☐ ☐
17. Are critical lift plans available? ☐ ☐ ☐
18. Is minimum clearance distance for high voltage lines posted at the operator's position? ☐ ☐ ☐
19. Do older lattice boom cranes with anti-two-block warning devices in lieu of anti-two-block devices have manually activated friction brakes? ☐ ☐ ☐
20. Is the slow-moving emblem used on all vehicles which by design move at 25 mi/h or less on public roads? ☐ ☐ ☐
21. Are all vehicles which will be parked or moving slower than normal traffic on haul roads equipped with a yellow flashing light or flasher visible from all directions? ☐ ☐ ☐
22. Is all equipment to be operated on public roads provided with
    - Headlights? ☐ ☐ ☐
    - Brake lights? ☐ ☐ ☐
    - Taillights? ☐ ☐ ☐
    - Backup lights? ☐ ☐ ☐
    - Front and rear turn signals? ☐ ☐ ☐
23. Are seat and seat belts provided for the operator and each rider on equipment? ☐ ☐ ☐
24. Is all equipment with windshields equipped with powered wipers and defogging or defrosting devices? ☐ ☐ ☐
25. Is the glass in the windshield or other windows clear and unbroken to provide adequate protection and visibility for the operator? ☐ ☐ ☐
26. Is all equipment equipped with adequate service brake system and emergency brake system? ☐ ☐ ☐
27. Are areas on equipment where employees walk or climb equipped with platforms, foot walks, steps, handholds, guardrails, toe boards, and nonslip surfaces? ☐ ☐ ☐

Inspections

28. Is self-propelled equipment equipped with automatic, audible, reverse signal alarms? ☐ ☐ ☐

29. Is there a record of manufacturer's approval of any modification of equipment which affects its capacity or safe operation? ☐ ☐ ☐

30. Are truck and crawler cranes, mounted on a barge or pontoon, attached by means of a tie-down system with some slack? ☐ ☐ ☐

Note: Movement during lifting is not permitted.

31. Have the following conditions been met for land cranes mounted on barges or pontoons:
    - Have load ratings been modified to reflect the increased loading from list, trim, wave, and wind action? ☐ ☐ ☐
    - Are all deck surfaces above the water? ☐ ☐ ☐
    - Is the entire bottom area of the barge or pontoon submerged? ☐ ☐ ☐
    - Are tie downs available? ☐ ☐ ☐
    - Are cranes blocked and secured? ☐ ☐ ☐

32. Are all belts, gears, shafts, spindles, drums, flywheels, or other rotating parts of equipment guarded where there is a potential for exposure to workers? ☐ ☐ ☐

33. Is the area where the crane is to work level, firm, and secured? ☐ ☐ ☐

34. Is a dry chemical or carbon dioxide fire extinguisher rated at least 5-BC on the crane? ☐ ☐ ☐

35. Are trucks, for truck-mounted cranes, equipped with a working reverse signal alarm? ☐ ☐ ☐

36. Is a signal person provided where there is danger from swinging loads, buckets, booms, etc.? ☐ ☐ ☐

37. Is there adequate clearance from overhead structures and electrical sources for the crane to be operated safely? ☐ ☐ ☐

38. Is there adequate lighting for night operations? ☐ ☐ ☐

39. Has the boom stop test on cable-supported booms been performed? ☐ ☐ ☐

40. Is the boom disengaging device functioning as required? ☐ ☐ ☐

41. Has all rigging and wire rope been inspected? ☐ ☐ ☐
42. Have performance load tests been in accordance with manufacturer's instructions? ☐ ☐ ☐
43. Are all environmental considerations met? ☐ ☐ ☐
44. Are communications provided as required? ☐ ☐ ☐

Source: USACE-EM 385 1-1, Department of the Army.

PORTAL, TOWER, AND PILLAR CRANES YES NO N/A

1. Are the following available:
    - Manufacture's written erection instructions? ☐ ☐ ☐
    - Listing of the weight of each component? ☐ ☐ ☐
    - An activity hazard analysis for the erection? ☐ ☐ ☐
    - Does the activity hazard analysis contain
        - Location of crane and adjacent structures? ☐ ☐ ☐
        - Foundation design and construction requirements? ☐ ☐ ☐
        - Clearance and bracing requirements? ☐ ☐ ☐
2. Is there a boom angle indicator within the operator's view? ☐ ☐ ☐
3. Are luffing jib cranes equipped with
    - Shock-absorbing jib stops? ☐ ☐ ☐
    - Jib hoist limit switch? ☐ ☐ ☐
    - Jib angle indicator visible to operator? ☐ ☐ ☐
4. If used, do rail clamps have slack between the point of attachment to the rail and the end fastened to the crane? ☐ ☐ ☐
5. Are the following with the crane at all times:
    - The manufacturer's operating manual? ☐ ☐ ☐
    - The load-rating chart? ☐ ☐ ☐
    - The crane's logbook documenting use, maintenance, inspections, and tests? ☐ ☐ ☐
    - The operating manual for crane operational aids used on the crane? ☐ ☐ ☐
6. Are all crane and derrick inspections performed by a qualified person? ☐ ☐ ☐
7. Are the following done on the project site:
    - Completed periodic inspection report prior to initial work? ☐ ☐ ☐

58   Inspections

- Preoperational checklist used for daily inspections? ☐ ☐ ☐
- Written reports of the operational performance tests? ☐ ☐ ☐
- Written reports of the load performance tests? ☐ ☐ ☐

8. Is every crane operator certified by a physician to be physically qualified to perform work? ☐ ☐ ☐
9. Are all operators qualified by written and practical exam or by appropriate licensing agency for the type of crane they are to operate? ☐ ☐ ☐
10. Is an activity hazard analysis for setup and set-down available? ☐ ☐ ☐
11. Is an activity hazard analysis for setup and set-down available? ☐ ☐ ☐
12. Are there at least three wraps of cable on the drum? ☐ ☐ ☐
13. Are the hoisting ropes installed IAW the manufacturer's recommendations? ☐ ☐ ☐
14. Is the record of manufacturer's approval of any modification of equipment which affects its capacity or safe operation? ☐ ☐ ☐

Source: USACE-EM (385-1-1), Department of the Army.

SAFE LIFTING OPERATIONS   YES   NO   N/A

Before Lifting Load

1. Has a qualified rigger been appointed to carry out rigging and signaling activities on the worksite? ☐ ☐ ☐
   - Has the appointed rigger successfully undergone training in rigging? ☐ ☐ ☐
2. Has the rigger checked the lifting gears or appliances such as chain blocks, wire ropes, shackles, eyebolts, and others for
   - Visible defects? ☐ ☐ ☐
   - Maximum safety working load? ☐ ☐ ☐
   - Date of last test? ☐ ☐ ☐
   - Current color coding? ☐ ☐ ☐
3. Are there any visible defects in the welded eye piece or lifting lug of the load? ☐ ☐ ☐
4. Has the weight of load to be carried been confirmed that it is below the safe working load (SWL) of the lifting gear or appliances? ☐ ☐ ☐

5. Is the load, including all loose items and lifting attachments, properly secured? ☐ ☐ ☐
6. Are pads in place, in areas where the wire ropes are bent around sharp edges? ☐ ☐ ☐
7. Is there one trained person around to give signals? ☐ ☐ ☐
8. Are all appropriate tag lines attached to the load? ☐ ☐ ☐
9. Has the weight of the load that is to be lifted been ascertained? ☐ ☐ ☐
   - Has the crane operator been informed of the weight of the load? ☐ ☐ ☐

During Lifting Load

10. Is the load properly balanced? ☐ ☐ ☐
11. Has the load been prevented from swinging? ☐ ☐ ☐
12. Are the loose chain or wire rope slings properly secured? ☐ ☐ ☐
13. Have the workers standing or working below the suspended load being cleared away? ☐ ☐ ☐
14. Are operators PROHIBITED to exceed Crane Load Ratings? ☐ ☐ ☐
15. Are operators PROHIBITED to tip the crane to determine allowable loads? ☐ ☐ ☐

Note: When either radius or boom length, or both, are between listed values, ALWAYS use the smaller of the two listed load ratings.

16. Are operators PROHIBITED to operate at longer radii than those listed on the applicable load rating chart? ☐ ☐ ☐

Note: The boom angles shown on the Capacity Chart give an approximation of the operating radius for a specified boom length. The boom angle, before loading, should be greater to account for boom deflection. It may be necessary to retract the boom if maximum boom angle is insufficient to maintain rated radius.

17. Are power telescoping boom sections ALWAYS extended equally? ☐ ☐ ☐

Note: Rated loads include the weight of hook block, slings, and auxiliary lifting devices. Their weights shall be subtracted from the listed rated load to obtain the net load that can be lifted. When lifting over the jib, the weight of any hook block, slings, and auxiliary lifting devices at the boom head must be added to the load. When jibs are erected but unused add two times the weight of any hook block, slings, and auxiliary lifting devices at the jib head to the load.

18. Are operators PROHIBITED to exceed rated ☐ ☐ ☐
 loads—85% on outriggers or 75% on tires of the tipping load as determined by SAE Crane Stability Test Code J765a?

Note: Rated loads are based on freely suspended loads. No attempt shall be made to drag a load horizontally on the ground in any direction.

19. Are crane operation ratings reduced to allow for ☐ ☐ ☐
 adverse job conditions, such as soft or uneven ground, out-of-level conditions, high winds, side loads, pendulum action, jerking or sudden stopping of loads, hazardous conditions, experience of personnel, two machine lifts, traveling with loads, electric wires, etc.?

Note: Side pull on boom or jib is HAZARDOUS. Derating of the crane's lifting capacity is required when wind speed exceeds 20 mi/h. The center of the lifted load must never be allowed to move more than 3 ft (91.44 cm) off the center line of the base boom section due to the effects of wind, inertia, or any combination of the two.

"Use 2 ft (61.91 cm) off the center line of the base boom for a two-section boom, 3 ft (91.44 cm) for a three-section boom, or 4 ft (121.92 cm) for a four-section boom."

Note: The maximum load that can be telescoped is not definable because of variations in loadings and crane maintenance, but it is permissible to attempt retraction and extension if load ratings are not exceeded. Load ratings are dependent upon the crane being maintained according to manufacturer's specifications.

20. Are load handling devices, including hooks and ☐ ☐ ☐
 hook blocks, kept away from boom head at all times?
21. Are truck crane operators PROHIBITED to lift with ☐ ☐ ☐
 outrigger beams positioned between the fully extended and intermediate (pinned) positions?
22. Are operators PROHIBITED to use truck cranes not ☐ ☐ ☐
 equipped with equalizing (bogie) beams between the rear axles for lifting "on tires"?

Note: Operators using truck cranes equipped with equalizing beams and rear air suspension should "dump" the air before lifting "on tires."

When Lowering Load

23. Is the resting place for the load suitable? ☐ ☐ ☐
    - Upon resting, is the load stable? ☐ ☐ ☐
24. Is the chain or wire rope sling slackened before attempting to remove it? ☐ ☐ ☐
25. Are the shackled pins properly secured after removing the chain or wire rope sling? ☐ ☐ ☐

Note: Upon completion of work, MAKE SURE that all lifting gear or accessories are kept properly secured.

Maintenance guidelines

Use a good checklist, when performing preventive maintenance to make sure all the work items are performed, and make notes of any issues that should be checked into further.

- Establish a crane and derrick maintenance inspection program based on crane manufacturer's recommendations—and have a competent person inspect the equipment before each use and during its use.
- Take the following precautions before making adjustments and repairs to cranes:
  - Run the crane to a location where it will cause the least interference with other cranes and operations.
  - Ensure that all controllers are at the off position.
  - Open and lock the main or emergency switch in the open position.
  - Place warning or "out of order" signs on the crane and on the hook where they are visible to operators and riggers.
  - Provide rail stops or other suitable means to prevent interference from other cranes in operation on the same runway.
- Allow only a designated competent person to make any repairs or replacements, as needed, for safe operation.
- Correct any disclosed unsafe conditions by testing before resuming crane operations.
- PROHIBIT operation of a crane until all guards have been reinstalled, safety devices reactivated, and maintenance equipment removed.

Note: A documented annual inspection log MUST be kept with the crane at all times. Boom cable installation documents must be readily available as well.

Overhead and gantry cranes

- Establish a preventive maintenance program, based on the crane manufacturer's recommendations, for all overhead and gantry cranes.
- Operator MUST perform a pre-use inspection (once per shift) for the items listed as follows and remove from service any equipment found to be unsatisfactory:
- Check operating mechanism for proper operation and adjustment and note any unusual sounds or noise due to chain binding or bearing squeal
  - Limit switch—Test upper-limit switch. If the hoist has a lower-limit switch, test it with no load before lowering any load that could bring the lower-limit switch into operation.
  - Air or hydraulic systems—Check for leaks (as applicable) all along the air or hydraulic system, including tanks, valves, pumps, and lines (visual inspection from floor level only).
  - Hoist braking—Confirm that the brakes are functioning.
  - Hooks and hook latches—Check for: excessive throat opening, bent or twisted elements, and sticky swivel or rough surfaces. Check latches (if present) for damaged spring and bent or missing hardware. Check self-locking hooks (if present) for proper operation and locking.
  - Hoist ropes and end connections—Check for rope distortions, such as kinking, crushing, upstanding, birdcaging, main strand displacement or core protrusion, corrosion, and broken or cut strands. Note number, distribution, and type of visible broken wires.
  - Spooling—Check ropes for proper spooling on drum(s) and sheave(s).
  - Festoon or trolley wiring—Ensure, when applicable, that wire collects and moves freely.
  - Bridge—Look for loose items or obstructions (visual inspection from floor level only).
- MAKE SURE overhead crane stops limit travel of the trolley.
- Ensure provision of
  - Bridge and trolley bumpers or equivalent automatic devices.
  - Bridge trucks have tail sweeps.

Crawler and mobile cranes

- Visually examine the crane structure for any deformed, cracked, or corroded members in the structure and the boom.
- Check for
  - Loose bolts or rivets
  - Excessive wear on brake and clutch system parts

- Deformed wedges
- Defective cotter keys, pins, and guardrails
- Deterioration or leakage in air or hydraulic systems
- Poor adjustment or excessive wear of all control mechanisms
- Accuracy of marking on the load/radius indicator over the full range
- Mechanical components in good working order (gearbox, hydraulics, etc.)

- Check hydraulic systems.
  - Deterioration or leakage in air or hydraulic systems
  - Safe and effective operation on hoses, pumps, and motors
  - Levels of fluid
  - Air cleaners for replacement or cleaning
- Check control mechanisms and monitoring devices.
  - Control mechanisms—cables, brakes, and levers for poor adjustment or excessive wear
  - Accuracy of marking on the load/radius indicator over the full range
  - Load moment indicator (LMI), boom angle, boom length indicator, and anti-two-block (ATB) system, according to manufacturer's manual
- Check components.
  - Lights—burned or broken.
  - Wheels or tracks—worn.
  - Brakes—shoe wear.
  - Bridge bumpers and trolley end stop—loose, missing, improper placement.
  - Controllers and collector shoes or bars—worn, pitted, loose, broke, or faulty operation.
  - Foot walk—condition of the boards, railings, and ladders.
  - Gears—lack of lubrication or foreign material in gear teeth (indicated by grinding or squealing).
  - Oil—inspect ONLY after opening and locking out the main switch.

Note: Before closing the main switch, MAKE SURE that all controllers are in the "off" position.

- Check fire extinguisher—MUST be in the crane cab.
- Check lower crane boom to unload sheaves.
- Unwind all wire rope from the hoist drum to expose all parts of a rope, making sure that the rope does not rewind in the reverse direction.
- Inspect sheaves, sockets, dead ends, thimble joints, and all wire rope hardware.

- Check sheaves, during rope changes, for worn bearings, broken flanges, proper groove size, smoothness, and contour.
- Inspect all parts of the cable, cleaning wire rope only as required to complete an inspection.

Note: Excessive removal of lubricant will lead to damage.

- Re-lubricate rope to prevent corrosion, wear, friction, and drying out of the core.
- Check for ropes that may have been operated dry (unlubricated) and replace dry ropes.

Note: There may be hidden damage that is not detected by visual inspection.

- Compare rope length and diameter with the original dimensions.

Note: Lengthening, accompanied by diameter reduction, is often an indication of interior core defects.

- Establish a schedule of rope replacement to change wire rope before it breaks.

Note: Periodic replacements DO NOT take the place of inspections.

- Reduce the time between replacement, if rope breaks or inspections reveal abnormal wire breakage or defects.
- DO NOT make wire rope slings from used wire rope.

Inspections

To help determine when the crane is safe to operate, the operator who walks around the crane looking for defects or problem areas performs frequent inspections at the start of each shift. Components that have a direct bearing on the safety of the crane and whose status can change from day to day with use MUST be inspected daily, and when possible, observed during operation for any defects that could affect safe operation.

All inspections should be conducted using the manufacturer's maintenance and inspection records, forms/checklist, or equivalent that set forth specific inspection, operation, and maintenance criteria for each mobile crane and its lifting capacity. A typical inspections can include

- Complete inspection of the mechanical, structural, electrical, and safety systems as well as of the wire ropes and chains. (Inspection includes, but is not limited to, the requirements of all governmental regulating agencies and local jurisdictional entities.)

- Nondestructive testing of the load hooks for cracks and visual inspection for distortions.
- Inspection of all the structural load-bearing members, including the sheaves.
- Checking of overhead crane girders, rails, and columns to ensure structural integrity.
- Performance of dynamic/static load tests, if required.
- Rechecking, while under load, for possible defects such as distortions, cracks, loose bolts.
- Checking of brakes, clutches, sheaves, and wire rope assemblies.
- Conducting of, after the load test, operational tests to re-ensure that the unit is functioning properly.
- Reexamination of all safety devices.
- Setting up of a complete equipment file on each unit containing a daily, weekly, and monthly check sheet, and results of the initial inspection.

Upon completion of an inspection, a report MUST be prepared that lists any deficiencies to be corrected. Equipment that passes inspection MUST be affixed with a label indicating the inspection date, name of the inspector, and other relevant information.

In-service inspections can ascertain the integrity of these potentially dangerous machines and devices.

Routine inspection of the equipment, in accordance with all-applicable regulations, standards, and good engineering practice, ensures safe and proper working capabilities and conditions of

- Material-handling devices—Cranes, derricks, forklifts, helicopters, jacks, slings, shackles, hooks, etc.
- Personnel handling devices—Hoists, elevators, cable cars, etc.

Once any piece of lifting or hoisting equipment has been found to be in safe and proper working condition, documents of certification can be issued, in accordance with government regulations and local laws.

An annual inspection of the lifting and hoisting machinery MUST be made by a competent person—with records kept of the dates and results of each inspection. All crawler, truck, or locomotive cranes in use must meet the requirements as prescribed in the ANSI B30.5-1968, Safety Code for crawler, locomotive, and truck cranes.

MAINTENANCE INSPECTIONS

	YES	NO	N/A

1. Has a preventive maintenance program been established, based on crane manufacturer's recommendations? ☐ ☐ ☐

2. Are the following precautions taken before adjustments and repairs are started on cranes:
   - Run the crane to a location where it will cause the least interference with other cranes and operations? ☐ ☐ ☐
   - Provide rail stops or other suitable means to prevent interference from other cranes in operation on the same tracks? ☐ ☐ ☐
   - Ensure that all controllers are at the off position? ☐ ☐ ☐
   - Open and lock the main or emergency switch in the open position? ☐ ☐ ☐
   - Place warning or OUT OF ORDER signs on the crane and on the hook where they are visible from the floor? ☐ ☐ ☐
   - Lower hoist to unload rope sheaves? ☐ ☐ ☐
   - Unwind all wire rope from the hoist drum to expose all parts of a rope, making sure that the rope does not rewind in the reverse direction? ☐ ☐ ☐
   - Inspect sheaves, sockets, dead ends, thimble joints, and all wire rope hardware? ☐ ☐ ☐
   - Are sheave bearings worn? ☐ ☐ ☐
   - Are flanges broken? ☐ ☐ ☐
   - Do grooves have proper size, smoothness, and contour? ☐ ☐ ☐

3. Are all parts of the cable inspected, cleaning wire rope only as required to complete an inspection? ☐ ☐ ☐

Note: Excessive removal of lubrication will lead to damage.

4. Is rope re-lubricated to prevent corrosion, wear, friction, and drying out of the core? ☐ ☐ ☐

5. Have ropes been operated dry (unlubricated)? ☐ ☐ ☐

Note: Replace dry ropes. There may be hidden damage that is not detected by visual inspection.

6. Does the rope length and diameter match the original dimensions? ☐ ☐ ☐

Note: Lengthening accompanied by diameter reduction is often an indication of interior core defects.

7. Is the crane structure deformed or cracked? ☐ ☐ ☐
   - Are members in the structure and boom corroded? ☐ ☐ ☐
   - Are there loose bolts or rivets? ☐ ☐ ☐
   - Is there excessive wear on brake and clutch system parts? ☐ ☐ ☐
8. Is there deterioration or leakage in air or hydraulic systems? ☐ ☐ ☐
9. Are any control mechanisms poorly adjusted? ☐ ☐ ☐
   - Do they exhibit excessive wear? ☐ ☐ ☐
10. Are markings on the load/radius indicator accurate over their full range? ☐ ☐ ☐
11. Is there a schedule of rope replacement established to change wire rope before it breaks? ☐ ☐ ☐

Note: Periodic replacements do not take the place of inspections; if rope breaks or inspections reveal abnormal wire breakage or defects, reduce the time between replacement. DO NOT make wire rope slings from used wire rope.

12. Are operators PROHIBITED to operate the crane until all guards have been reinstalled, safety devices reactivated, and maintenance equipment removed? ☐ ☐ ☐
13. Are unsafe conditions disclosed by testing corrected before resuming crane operations? ☐ ☐ ☐
14. Are only designated personnel allowed to perform crane adjustments and repairs? ☐ ☐ ☐
15. Are repairs or replacements, as needed, promptly provided for safe operation? ☐ ☐ ☐

SAFETY INSPECTIONS   YES   NO   N/A

1. Are cranes visually inspected for defective components prior to the start of any work shift? ☐ ☐ ☐
2. Are all electrically operated cranes effectively grounded? ☐ ☐ ☐
3. Is a crane preventive maintenance program established? ☐ ☐ ☐
4. Is the load chart clearly visible to the operator? ☐ ☐ ☐
5. Are all operators trained and provided with the operator's manual for the particular crane being operated? ☐ ☐ ☐

Inspections

6. Have operators of construction industry cranes of 5-ton (4.54-metric ton) or greater capacity been issued a valid operator's card? ☐ ☐ ☐
7. Are operating controls clearly identified? ☐ ☐ ☐
8. Is a fire extinguisher provided at the operator's station? ☐ ☐ ☐
9. Is the rated capacity visibly marked on each crane? ☐ ☐ ☐
10. Is an audible warning device mounted on each crane? ☐ ☐ ☐
11. Is sufficient lighting provided for the operator to perform the work safely? ☐ ☐ ☐
12. Are cranes with booms that could fall backward equipped with boom stops? ☐ ☐ ☐
13. Does each crane have a certificate indicating that required testing and examinations have been performed? ☐ ☐ ☐
14. Are crane inspection and maintenance records maintained and available for inspection? ☐ ☐ ☐

Main Boom, Jib Boom, Boom Extension

15. Are boom jibs, or extensions, cracked or corroded? ☐ ☐ ☐
16. Are bolts and rivets tight? ☐ ☐ ☐
17. Is certification that repaired boom members meet manufacturer's original design standard documented? ☐ ☐ ☐

Note: Noncertified repaired members MUST NOT be used until recertified.

Load Hooks and Hook Blocks

18. Are hooks and blocks permanently labeled with rated capacity? ☐ ☐ ☐
19. Are hooks and blocks counterweighted to the weight of the overhaul line from highest hook position? ☐ ☐ ☐
20. Do hooks must have cracks or throat openings more than 15% of normal? ☐ ☐ ☐
21. Are hooks twisted off center more than 10 degrees from the longitudinal axis? ☐ ☐ ☐
22. Are all hooks used to hoist personnel equipped with effective positive safety catches, especially on hydraulic cranes? ☐ ☐ ☐

Hydraulic Hoses, Fittings and Tubing

23. Are flexible hoses sound—and showing no signs of leaking at the surface or its junction with the metal and couplings? ☐ ☐ ☐
24. Do hoses show blistering or abnormal deformation to the outer covering? ☐ ☐ ☐
25. Are there any leaks at threaded or clamped joints that cannot be eliminated by normal tightening or recommended procedures? ☐ ☐ ☐
26. Is there any evidence of excessive abrasion or scrubbing on the outer surfaces of hoses, rigid tubing, or hydraulic fittings? ☐ ☐ ☐

Outriggers

27. Are outrigger numbers, locations, types, and type of control in accordance with the manufacturer's specifications? ☐ ☐ ☐
28. Are outriggers visible to the operator or a signal person during extension or setting? ☐ ☐ ☐

Note: Outriggers are designed and operated to relieve all weight from wheels or tracks within the boundaries of the outriggers; if not, the manufacturer's specifications and operating procedures MUST be clearly defined.

Load-Rating Chart

29. Is a durable rating chart(s) with legible letters and figures attached to the crane in a location accessible to the operator while at the controls? ☐ ☐ ☐
30. Does the chart(s) contain the following:
    - Full and complete range of manufacturer's crane loading ratings at all stated operating radii? ☐ ☐ ☐
    - Optional equipment on the crane such as outriggers and extra counterweight which affect ratings? ☐ ☐ ☐
    - Work area chart for which capacities are listed in the load-rating chart, i.e., over side, over rear, over front? ☐ ☐ ☐
    - Weights of auxiliary equipment, i.e., load block, jibs, boom extensions? ☐ ☐ ☐
    - Clearly distinguishable list of ratings based on structural, hydraulic, or other factors rather than stability? ☐ ☐ ☐

- List of no-load work areas? ☐ ☐ ☐
- Description of hoist line reeving requirements on the chart or in operator's manual? ☐ ☐ ☐

Wire Rope

31. Does the main hoist and auxiliary wire rope inspection include examining for
    - Broken wires? ☐ ☐ ☐
    - Excessive wear? ☐ ☐ ☐
    - External damage from crushing, kinking, cutting, or corrosion? ☐ ☐ ☐

Cab

32. Does the cab contain all crane function controls—in addition to mechanical boom angle indicators, electric wipers, dash lights, warning lights and buzzers, fire extinguishers, seat belts, horn, and clear unbroken glass? ☐ ☐ ☐

Braking Systems

33. Are truck cranes and self-propelled cranes mounted on rubber-tired chassis? ☐ ☐ ☐
    - Or, are frames equipped with a service brake system, secondary stopping emergency brake system, and a parking brake system? ☐ ☐ ☐

Note: Not applicable, if the operator can show written evidence that such systems were not required by the standards or regulations in force at the date of manufacture and are not available from the manufacturer.

34. Has the braking system been inspected and tested and found to be in conformance with applicable requirements? ☐ ☐ ☐

35. Are crawlers provided with brakes or other locking devices that effectively hold the machine stationary on level grade during the working cycle? ☐ ☐ ☐
    - Is the braking system capable of stopping and holding the machine on the maximum grade recommended for travel? ☐ ☐ ☐

Note: The brakes or locks are arranged to engage or remain engaged in event of loss of operating pressure or power.

Turntable / Crane Body

36. Is the rotation point of a crane's gears and rollers free of damage and wear and properly adjusted? ☐ ☐ ☐
    - Are the components securely locked and free of cracks or damage? ☐ ☐ ☐
    - Is the swing locking mechanism functional (pawl, pin) and operated in the cab? ☐ ☐ ☐

Counterweight

37. Is the counterweight approved and installed according to manufacturer's specifications with attachment points secured? ☐ ☐ ☐

Sources: OSHA 29 CFR 1926.550 and ANSI B30.5 Standards.

Inspection schedules

Initial inspection. Prior to initial use, all new and altered cranes ARE to be inspected to ensure compliance with OSHA provisions 29 CFR 1910.180 and 29 CFR 1926.

Regular inspections. Inspection procedure for cranes in regular service is divided into two general classifications based upon the intervals at which inspection should be performed. The intervals in turn are dependent upon the nature of the critical components of the crane and the degree of their exposure to wear, deterioration, or malfunction.

- Frequent inspections—Daily to monthly intervals
- Periodic inspections—1- to 12-month intervals, or as specifically recommended by the manufacturer

	YES	NO	N/A
1. Are all new and altered cranes inspected before first use to ensure compliance with regulations?	☐	☐	☐

Note: Crane inspections are broken into frequent inspections of daily to monthly intervals and periodic inspections of 1- to 12-month intervals.

72   Inspections

2. Do frequent inspections include the following:
   - All functional operating mechanisms, for maladjustments interfering with proper operation (daily)? ☐ ☐ ☐
   - Deterioration or leakage in lines, tanks, valves, drain pumps, and other parts of air or hydraulic systems (daily)? ☐ ☐ ☐
   - Hooks with deformations or cracks:
     - Visual inspection (daily)? ☐ ☐ ☐
     - Signed reports (monthly)? ☐ ☐ ☐
   - Hoist chains (including end connections), for excessive wear, twists, distorted links interfering with proper function or stretch beyond manufacturer's recommendations—visual inspection (daily) and with signed reports (monthly)? ☐ ☐ ☐
   - All functional operating mechanisms, for excessive wear of components? ☐ ☐ ☐
   - Rope reeving, for noncompliance with manufacturer's recommendations? ☐ ☐ ☐

3. Are periodic inspections complete, and do they include the requirements of frequent inspections as well as checking for the following:
   - Deformed, cracked, or corroded members? ☐ ☐ ☐
   - Loose bolts or rivets? ☐ ☐ ☐
   - Cracked or worn sheaves and drums? ☐ ☐ ☐
   - Worn, cracked, or distorted parts such as pins, bearings, shafts, gears, rollers, and locking and clamping devices? ☐ ☐ ☐
   - Excessive wear on brake system parts, linings, pawls, and ratchets? ☐ ☐ ☐
   - Load, wind, and other indicators over the full range for any inaccuracies? ☐ ☐ ☐
   - Gasoline, diesel, electric, or other power plants for improper performance or noncompliance with applicable safety requirements? ☐ ☐ ☐
   - Excessive wear of chain drive sprockets and excessive chain stretch? ☐ ☐ ☐
   - Electrical apparatus for signs of pitting or any deterioration of controller contacts, limit switches, and push-button stations? ☐ ☐ ☐

- Cranes that have been idle for at least 1 month, but less than 6 months, per requirements for frequent, periodic rope inspections? ☐ ☐ ☐
- Cranes idle for more than 6 months, per requirements for frequent, periodic rope inspections? ☐ ☐ ☐
- Standby cranes at least semiannually, in accordance with frequent, periodic rope inspection maintenance? ☐ ☐ ☐

FREQUENT INSPECTIONS—DAILY TO MONTHLY INTERVALS YES NO N/A

1. Are inspections of cranes or derricks made at the start of each shift that include, but are not limited to,
   - Guarding of all exposed moving parts? ☐ ☐ ☐

Note: A removed guard may indicate that a mechanic is still working on part of the crane.

   - Any defects in each component of the crane used in lifting, swinging, or lowering the load or boom that might result in unsafe operation? ☐ ☐ ☐
   - All wire rope (including standing ropes), sheaves, drums rigging, hardware, and attachments? ☐ ☐ ☐
   - Hooks that are deformed or cracked? ☐ ☐ ☐

Note: Hooks with cracks, excessive throat openings of 15%, or hook twists of 10 degrees or more, MUST be removed from service.

   - Swivels free to rotate? ☐ ☐ ☐
   - Boom and jib for straightness and any evidence of physical damage, such as cracking, bending, or any other deformation of the welds? ☐ ☐ ☐
   - Corrosion under any attachments that are connected to the chords and lacing? ☐ ☐ ☐

Note: Look carefully for cracking or flaking of paint, an indication of possible fatigue of the metal that often precedes a failure.

2. Is lacing on lattice booms bent? ☐ ☐ ☐

Note: If they are kinked or bent, the main chord can lose substantial support in that area. When lacing is bent, the ends also tend to draw together, pulling the main chords out of shape. This precaution is especially important on tubular booms where every component must be straight and free from any dents. DO NOT attempt to straighten these members by hammering or heating them and drawing them out. They must be cut out and replaced with lacing to the manufacturer's specifications, procedures, and approval.

3. Do the tires have cuts, tears, or breaks? ☐ ☐ ☐
   - Are they properly inflated? ☐ ☐ ☐
4. Does the crane have fluid leak—air or hydraulic, or both? ☐ ☐ ☐
5. Is the crane properly lubricated? ☐ ☐ ☐
6. Are fuel, lubricating oil, coolant, and hydraulic oil reservoirs filled to proper levels? ☐ ☐ ☐
7. Is the crane equipped with a fully charged fire extinguisher? ☐ ☐ ☐
   - Does the operator know how to use it? ☐ ☐ ☐
8. Are all functional operating mechanisms such as sheaves, drums, brakes, locking mechanisms, hooks, the boom, jib, hook rollers brackets, outrigger components, limit switches, safety devices, hydraulic cylinders, instruments, and lights in working order? ☐ ☐ ☐
9. Do turntable connections have weld cracks and loose or missing bolts? ☐ ☐ ☐

Note: If bolts are loose, there is a good chance that they have been stretched.

10. Are the beams or the cylinders of outriggers distorted? ☐ ☐ ☐
    - Are any welds cracked? ☐ ☐ ☐
    - Do both the beams and cylinders extend and retract smoothly and hold the load? ☐ ☐ ☐
    - Check the condition of the floats, and are they securely attached? ☐ ☐ ☐
11. Are all brakes and clutches properly adjusted and operational? ☐ ☐ ☐
12. Are boom hoist lockout and other operator aids, such as anti-two-block (ATB) devices and load moment indicators (LMI), operating properly and calibrated correctly? ☐ ☐ ☐
13. Do all gauges and warning lights, while engine is running, function properly for proper readings and operation of all controls? ☐ ☐ ☐
14. Is there any broken or cracked glass that may affect the view of the operator? ☐ ☐ ☐

PERIODIC INSPECTIONS—1- TO 12-MONTH INTERVALS

The periodic inspection determines the need for repair or replacement of components to keep the machine in proper operating condition and includes those items listed for daily inspections as well as, but not limited to, structural defects, excessive wear, and hydraulic or air leaks.

	YES	NO	N/A
1. Is the entire crane inspected for structural damage?	☐	☐	☐

Note: Be careful to check for distortion or cracks in the main frame, outrigger assemblies, and structural attachments of the upper works to the carrier.

	YES	NO	N/A
2. Are all welded connections inspected for cracks?	☐	☐	☐
▪ Main chords and lacings and other structural items—for paint flaking and cracking that may indicate potential failure, as well as for dents, bends, abrasions, and corrosion?	☐	☐	☐
▪ Hydraulic booms—for bending, side sway, or droop?	☐	☐	☐
3. Are there any deformed, cracked, or corroded members in load/stress bearing structure?	☐	☐	☐

Note: Magnetic particle or other suitable crack detecting inspection should be performed at least once each year by an inspection agency retained by the owner. Inspection reports should be requested and retained in the crane file.

	YES	NO	N/A
4. Are sheaves and drums cracked or worn?	☐	☐	☐
5. Are there any worn, cracked, or distorted parts such as pins, bearings, shafts, gears, rollers, locking devices, hook roller brackets, removable outrigger attachments lugs, and welds?	☐	☐	☐
6. Is there excessive wear on brake and clutch system parts, linings, pawls, and ratchets?	☐	☐	☐
7. Are all load and boom angle indicators operating properly?	☐	☐	☐
▪ Are they calibrated correctly?	☐	☐	☐
8. Do all power plants operate properly?	☐	☐	☐
9. Is there excessive wear on drive sprockets or is there chain stretch, or both?	☐	☐	☐
10. Do steering, braking, and locking devices have correct action?	☐	☐	☐

11. Is the counterweight secure? ☐ ☐ ☐
12. Is the identification number permanently and legibly marked on jibs, blocks, equalizer beams, and all other accessories? ☐ ☐ ☐
13. Do hydraulic and pneumatic hoses, fittings, and tubing have any conditions, such as:
    - Evidence of oil or air leaks on the surfaces of flexible hoses or at the point at which the hose in question joins the metal end couplings? ☐ ☐ ☐
    - Abnormal deformation of the outer covering of hydraulic hose, including any enlargement, local or otherwise? ☐ ☐ ☐
    - Leakage at connections which cannot be eliminated by normal tightening? ☐ ☐ ☐
    - Evidence of abrasive wear that could have reduced the pressure-retaining capabilities of the hose or tube effected. The cause of the rubbing or abrasion must be immediately eliminated? ☐ ☐ ☐

Any deterioration of any system component requires immediate evaluation of whether further use would constitute a safety hazard that would require replacement of the part in question.

Inspection records, of the inspected crane, MUST be maintained monthly on critical items in use, such as brakes, crane hooks, and ropes. The records should include the date of inspection, the signature of the person who performed the inspection, and the serial number, or other identifier, and should be kept readily available for review.

These inspections begin with a general walk-around and observation of the overall crane setup and operation, followed by a specific inspection of items or components.

In general, when inspecting any crane, MAKE SURE to request for, and review, all inspection and maintenance documents for the crane being inspected, including the crane manufacturer's inspection and maintenance requirements. Please use the checklist below when conducting a walk-around inspection.

WALK-AROUND (START OF EACH SHIFT)   YES   NO   N/A

1. Do mechanical systems have any leaks or damage (oil, hydraulic, air), or structural deficiencies? ☐ ☐ ☐
2. Are crane controls properly marked? ☐ ☐ ☐
    - Any instruments damaged? ☐ ☐ ☐
    - Load charts properly displayed and legible? ☐ ☐ ☐

3. Does the operator, ground crew (riggers), or supervisor have any questions on load charts, rigging and load weight determinations, and capacities? ☐ ☐ ☐
4. Does the operator have adequate braking ability to stop when raising or lowering the boom/load? ☐ ☐ ☐

Note: Inspect, from the cab position, the running line or rope of the main hoist drum and secondary line or jib line, as the operator raises and lowers the boom/load line.

5. Are there any abnormal or defective conditions of boom sections, lacing, lifting components, anti-two-block devices, jib back stops, and the hook? ☐ ☐ ☐
6. Do outriggers on hydraulic cranes provide adequate stability in setting-up? ☐ ☐ ☐
7. Does cribbing on crawler cranes provide adequate stability in setting-up? ☐ ☐ ☐

Note: Request the crane to be rotated to check all clearances and overall stability.

Source: Crane Inspection & Certification Bureau

8. Are all exposed moving parts such as gears, chains, and reciprocating or rotating parts guarded or isolated? ☐ ☐ ☐
9. Are protection barriers provided for guarding the rear swing area? ☐ ☐ ☐
10. Are high-voltage warning signs, displaying restrictions and requirements, installed at the operator's station and at strategic locations on the crane? ☐ ☐ ☐
11. Are shock-absorbing or hydraulic-type boom stops installed in a manner to resist boom overturning? ☐ ☐ ☐
12. Are jib stops installed to resist overturning? ☐ ☐ ☐
13. Is a boom angle indicator, readable for the operator station, installed accurately to indicate boom angle? ☐ ☐ ☐
14. Is a boom hoist disconnect safety shutoff or hydraulic relief installed that automatically stops the boom hoist when the boom reaches a predetermined high angle? ☐ ☐ ☐

15. Are cranes with telescoping booms equipped with a ☐ ☐ ☐
    two-blocking damage prevention feature that has
    been tested on-site in accordance with manufacturer's
    requirements?

Note: All cranes, hydraulic and fixed boom used to hoist personnel, MUST be equipped with two-block devices on all hoist lines intended to be used in the operation. The anti-two-block device has automatic capabilities for controlling functions that may cause a two-blocking condition.

16. Are cranes, used to hoist personnel, equipped for ☐ ☐ ☐
    power-controlled lowering operation on all hoist
    lines?

Note: Check clutch, chains, and sprockets for wear.

17. Is a device or procedure for leveling the crane ☐ ☐ ☐
    provided?
18. Are sheave grooves smooth and free from surface ☐ ☐ ☐
    defects, cracks, or worn places that could cause rope
    damage?
    - Are flanges broken, cracked, or chipped? ☐ ☐ ☐
    - Does the bottom of the sheave groove form a ☐ ☐ ☐
      close fitting saddle for rope being used?
    - Are lower load blocks equipped with close fitting ☐ ☐ ☐
      guards?

Note: Almost every wire rope installation has one or more sheaves—ranging from traveling blocks with complicated reeving patterns to equalizing sheaves where only minimum rope movement is noticed.

Regulations

29 CFR 1910   Labor/Occupational Safety and Health Standards
   Subpart N   Materials Handling and Storage
     .178   Powered industrial trucks
     .179   Overhead and gantry cranes
     .180   Crawler locomotive and truck cranes
     .181   Derricks
29 CFR 1926   Labor/Safety and Health Regulations for Construction
   Subpart N   Material Hoists, Personnel Hoists, and Elevators
     .550   Cranes and derricks
     .552   Cranes, derricks, hoists, elevators, and conveyors
     .553   Base-mounted drum hoists
     .554   Overhead hoists

Standards (consensus)

ANSI
- B 30.2 Safety Code for Overhead and Gantry Cranes (1967)
- B 30.5 Rough Terrain Hydraulic Cranes
- B 30.5 Lattice-Boom, Crawler Crane

ANSI/ASSE
- A 10.28 Work Platforms Suspended from Cranes or Derricks (R 2004)

ASME
- B 30.2 Overhead and Gantry Cranes (R 2005) (Top Running Bridge/Single or Multiple Girder/Top Running Trolley Hoist)
- B 30.3 Construction Tower Cranes (R 2009)
- B 30.4 Portal, Tower, and Pedestal Cranes—Permanently Mounted Cranes (R 2003)
- B 30.5 Mobile and Locomotive Cranes (R 2007)
- B 30.6 Derricks (R 2003)
- B 30.8 Floating Cranes and Floating Derricks—Mounted on Barges (R 2004)
- B 30.11 Monorails and Underhung Cranes (R 2004)
- B 30.17 Overhead and Gantry Cranes (2006) (Top Running Bridge, Single Girder, Underhung Hoist)
- B 30.20 Standard for Design, Testing, and Appropriate Markings
- B 30.22 Articulating Boom Cranes (R 2005)
- B 30.26 Rigging Hardware
- B 30.9 Slings
- B 30.10 Hooks
- NOG-1 Overhead and Gantry Cranes
- NUM-1 Cranes, Monorails, and Hoists

ASME/ANSI
- B 30.1 General Purpose Portable Jacks
- B 30.5 Crawler, Locomotive and Truck Cranes
- Safety Code for Cranes, Derricks, and Hoists
- B 30.11 Monorails and Underhung Cranes (1988)
- B 30.17 Overhead and Gantry Cranes (1985) (Top Running Bridge, Single Girder, Underhung Hoist)

ASSE/SAFE
- A10.13 Safety Requirements for Steel Erection
- American National Standard for Construction and Demolition Operations
- A10.28 Safety Requirements for Work Platforms Suspended from Cranes or Derricks

American National Standard for Construction and Demolition Operations
A10.42 Safety Requirements for Rigging Qualifications and Responsibilities
American National Standard for Construction and Demolition Operations

CMAA
Standard No. 70
No. 74 Top Running and Under Running Single Girder, Electric Overhead Traveling Cranes

PCSA
No. 1 Mobile Power Crane (and Excavator) and Hydraulic Crane
No. 2 Mobile Hydraulic Crane
No. 4 Mobile Power Crane (and Excavator) and Hydraulic Crane

SAE
J376-95 Load-Indicating Devices in Lifting Crane Service
J765A Crane Load Stability Test Code
J874 Center of Gravity Test Code
J1063 Method of Test for Crane Structure

4.2 Forklifts

Guidelines

Operational. Heavy material handling that requires the use of forklifts and other powered industrial trucks is a potentially hazardous activity. To ensure the safety of such operations requires worker training and stringent controls for the use of such equipment, its procurement, maintenance, and inspection.

Design specifications and performance characteristics of forklifts define their capabilities and limitations under adverse conditions—thus, an operator must know its capabilities and limitations to operate a forklift efficiently.

Two major classifications of material-handling vehicles are generally used in various industries:

Rough-terrain forklifts—Internal combustion engines machines that can use diesel, gasoline, or propane fuel depending on manufacturer specifications. Equipped with big pneumatic tires having deep threads allowing the vehicle to navigate through the roughest areas without slipping or sliding to pick up and move raw materials, processed parts, finished products, tools, equipment, supplies, or maintenances items.

Industrial-powered trucks—Either electric-, gasoline-, or propane-powered having solid, semisolid, or pneumatic rubber tires—for use in factories, warehouses, and freight terminals, or on hard-surfaced outdoor storage areas, to perform most of the lifting and stacking involved in operations. There are many types and sizes for different jobs, named by the function they perform, such as high lift trucks, counterbalanced trucks, and rider trucks.

Forklift attachments—Whenever an attachment is used that could affect the capacity or safe operation of a forklift,

> Forklift manufacturer MUST approve the attachment
> Employer MUST MARK the forklift to show the new weight with attachment—as well as the forklift's maximum capacity at the highest elevation

Maintenance. Use a good checklist when performing preventive maintenance to make sure all the work items are performed, and make notes of any issues that should be checked into further.

- Perform regular preventive maintenance.
  - Check the horn, gauges, safety warning devices, cab lights, outside light, backup lights, and overhead guard for proper operation.
  - Check battery compartment, battery terminal, battery cables, and battery connectors.
  - Check cells for proper fill levels.
  - Apply protectant on the connectors to prevent corrosion.
- Drain forklift oil.
  - Change oil filter.
  - Change hydraulic filter, transmission filter, fuel filters, and air filter.
- Check all fluid levels and fill as necessary.
  - Check antifreeze in cooling system.
- Visually inspect drive axle (while underneath the machine).
  - Grease the drive axle (if there is a grease fitting).
- Inspect belts on forklift engine.
  - Check for cracks and frays on belts.
- Visually inspect forklift's exhaust system.
  - Check for cracks or holes in the system.
- Inspect tires for any cracks or signs of excessive wear.
  - Check air on inflate tire.
  - Check tire tread.
  - Check differential fluid.

- Inspect forklift operation systems.
  - Check clutch pedal.
  - Check steering cylinder and drag links.
  - Check for any leaks or bends in the system.
- Inspect the forklift hoist.
  - Check hydraulic lines for any leaks.
- Inspect the forks and mast.
  - Check for any visible cracks or chips.
  - Check for any bowing or bends in the forks.
- Lubricate all forklift fittings, including hoist mechanism and tie-rod ends.
- Check wheel bearings, when replacing forklift brakes.
- Wipe down the forklift, using an oil-based spray cleaner.
- ALWAYS follow proper safety procedures—to prevent injury or illness when doing maintenance on a forklift.
  - DO NOT make repairs in an area with a potentially flammable or combustible atmosphere.
  - ALWAYS make sure there is adequate ventilation to prevent accumulation of exhaust or gas fumes.
  - DO NOT use flammable solvent to clean a forklift; use a noncombustible (flash-point above 100°F) solvent.
  - NEVER get under a forklift supported only by a jack or under any part supported only by hydraulic pressure.
  - ALWAYS install jack stands or a secure block support.
  - REMOVE KEY or disconnect the battery while making repairs—to prevent the forklift from accidentally being started.

INSPECTIONS	YES	NO	N/A
1. Are forklifts inspected before each shift or before being used on each day?	☐	☐	☐
■ Are they visually checked for loose or broken parts—wheel lugs, oil leaks, etc.?	☐	☐	☐
2. Are tires or wheels damaged?	☐	☐	☐
■ Are they lubricated for smooth operation?	☐	☐	☐
3. Are bumpers and end stops in place and firmly secured?	☐	☐	☐
4. Is the annual inspection sticker current?	☐	☐	☐
5. Is the maximum weight load capacity posted for operator visibility?	☐	☐	☐

Lifting Equipment 83

6. Are all moving parts properly lubricated to move freely? ☐ ☐ ☐
7. Do welds and frame have any structural damage? ☐ ☐ ☐
8. Are the following equipment items checked for damage:
   - Name plates, markings, and load limits? ☐ ☐ ☐
   - Operating lights? ☐ ☐ ☐
   - Mast, carriage, and attachments? ☐ ☐ ☐
   (Missing bolts, unusual wear on chain guides, and inside of mast channels)
   - Forks and fork adjusting slides? ☐ ☐ ☐
   - Leaks under forklift? ☐ ☐ ☐
   - Seat and seat belt? ☐ ☐ ☐
   - Backrest? ☐ ☐ ☐
   - Tires/axles? ☐ ☐ ☐
   (Cuts, gouges, imbedded objects, and air pressure for pneumatic tires)
   - Hydraulic hoses? ☐ ☐ ☐
   - Fuel line leaks and damage? ☐ ☐ ☐
   - Exhaust system (sparks and leaks)? ☐ ☐ ☐
   - Belts and pulleys? ☐ ☐ ☐
   - Overhead guard (ROPS)? ☐ ☐ ☐
   - Fire extinguisher? ☐ ☐ ☐
9. Are the following power plant items checked for problems:
   - Battery connection? ☐ ☐ ☐
   - Battery level and amp gauge? ☐ ☐ ☐
   - Fuel tank and connections, damage/leaks? ☐ ☐ ☐
   - Fuel lines/nozzle/valves? ☐ ☐ ☐
10. Are the following systems checked for proper operation?
    - All gages/indicators? ☐ ☐ ☐
    - Overhead and warning lights? ☐ ☐ ☐
    - Windshield wipers (if equipped)? ☐ ☐ ☐
    - Engine oil level? ☐ ☐ ☐
    - Coolant level? ☐ ☐ ☐
    - Fuel level? ☐ ☐ ☐
    - Hydraulic controls and fuel level? ☐ ☐ ☐
    - Starting? ☐ ☐ ☐
    - Clutch or creeper control? ☐ ☐ ☐
    - Battery power? ☐ ☐ ☐
    - Lift operation? ☐ ☐ ☐

84   Inspections

- Tilt operation? ☐ ☐ ☐
- Horn or backup alarm (or both)? ☐ ☐ ☐
- Steering? ☐ ☐ ☐
- Parking and operational brakes? ☐ ☐ ☐

11. Is the forklift clean, free of dirt, excess oil, and grease? ☐ ☐ ☐

Note: Any items found to be defective require immediate notification of a supervisor, and the unit MUST be taken out of service until repaired.

Refer to OSHA Standard 1910.178 and ANSI Standard B56.1 for more detailed information.

INDUSTRIAL FORKLIFTS   YES   NO   N/A

1. Is there safe clearance for equipment through aisles and doorways? ☐ ☐ ☐
2. Are aisles designated, permanently marked, and kept clear to allow unhindered passage? ☐ ☐ ☐
3. Are motorized vehicles and mechanized equipment inspected daily or prior to use? ☐ ☐ ☐
4. Are vehicles shut off and brakes set prior to loading or unloading? ☐ ☐ ☐
5. Are containers of combustibles or flammables, when stacked while being moved, always separated by dunnage sufficient to provide stability? ☐ ☐ ☐
6. Are dock boards (bridge plates) used when loading or unloading operations are taking place between vehicles and docks? ☐ ☐ ☐
7. Are trucks and trailers secured from movement during loading and unloading operations? ☐ ☐ ☐
8. Are dock plates and loading ramps constructed and maintained with sufficient strength to support imposed loading? ☐ ☐ ☐
9. Are hand trucks maintained in safe operating condition? ☐ ☐ ☐
10. Are chutes equipped with sideboards of sufficient height to prevent the materials being handled from falling off? ☐ ☐ ☐
11. Are chutes and gravity roller sections firmly placed or secured to prevent displacement? ☐ ☐ ☐
12. At the delivery end of the rollers or chutes, are provisions made to break the movement of the handled materials? ☐ ☐ ☐

13. Are pallets usually inspected before being loaded or moved? ☐ ☐ ☐
14. Are hooks with safety latches or other arrangements used when hoisting materials so that slings or load attachments won't accidentally slip off the hoist hooks? ☐ ☐ ☐
15. Are securing chains, ropes, chocks, or slings adequate for the job to be performed? ☐ ☐ ☐
16. When hoisting material or equipment, are provisions made to ensure no one will be passing under the suspended loads? ☐ ☐ ☐
17. Are Material Safety Data Sheets available to employees handling hazardous substances? ☐ ☐ ☐

Regulations
  29 CFR 1910
    .178   Powered industrial trucks
  29 CFR 1926
    Subpart H   Materials Handling, Storage, Use, and Disposal
      .250   Rigging equipment for material handling
    Subpart O   Motor Vehicles, Mechanized Equipment, and Marine Operations
      .602   Material handling equipment
    Subpart W   Rollover Protective Structures, Overhead Protection
      1000   Rollover protective structures (ROPS) for material handling equipment

TRANSPORTING EMPLOYEES AND MATERIALS	YES	NO	N/A

1. Do employees who operate vehicles on public thoroughfares have valid operator licenses? ☐ ☐ ☐
2. Are motor vehicle drivers trained in defensive driving and proper use of the vehicle? ☐ ☐ ☐
3. When seven or more employees are regularly transported in a van, bus, or truck, is the operator's license appropriate for the class of vehicle being driven? ☐ ☐ ☐
4. Does each van, bus, or truck used regularly to transport employees have an adequate number of seats? ☐ ☐ ☐
5. Are seat belts provided and are employees required to use them? ☐ ☐ ☐

6. When employees are transported by truck, are provisions made to prevent their falling from the vehicle? ☐ ☐ ☐

7. Are vehicles used to transport employees equipped with lamps, brakes, horns, mirrors, windshields, and turn signals, and are they in good repair? ☐ ☐ ☐

8. Are transport vehicles provided with handrails, steps, stirrups, or similar devices, placed and arranged to allow employees to safely mount or dismount? ☐ ☐ ☐

9. Are employee transport vehicles equipped at all times with at least two reflective-type flares? ☐ ☐ ☐

10. Is a fully charged fire extinguisher, in good condition, with at least a 4-BC rating, maintained in each employee transport vehicle? ☐ ☐ ☐

11. When cutting tools or tools with sharp edges are carried in passenger compartments of employee transport vehicles, are they placed in closed boxes or containers that are secured in place? ☐ ☐ ☐

12. Are employees prohibited from riding on top of any load that could shift, topple, or otherwise become unstable? ☐ ☐ ☐

13. Are materials that could shift and enter the cab secured or barricaded? ☐ ☐ ☐

MATERIAL HANDLING AND STORAGE   YES   NO   N/A

1. Are all materials, which are stored in tiers, stacked, racked, blocked, interlocked, or otherwise secured to prevent sliding, falling, or collapse?
[29 CFR 1926.250 (a)(1)] ☐ ☐ ☐

2. Is the minimum safe load limit of floors within buildings and structures, in pounds per square foot, conspicuously posted in all storage areas?
[29 CFR 1926.250 (a)(2)] ☐ ☐ ☐

3. Are maximum safe loads always maintained?
[29 CFR 1926.120(a)(2)] ☐ ☐ ☐

4. Are aisles and passageways kept clear to provide for the free and safe movement of material handling equipment and people?
[29 CFR 1926.250(a)(3)] ☐ ☐ ☐

5. Are such areas kept in good repair?
[29 CFR 1926.250(a)(4)] ☐ ☐ ☐

Lifting Equipment 87

6. Where a difference in road or working level exists, are means such as ramps, blocking, or grading provided to ensure the safe movement of vehicles between two levels? ☐ ☐ ☐

7. Is material stored inside buildings under construction not placed within 6 ft (1.83 m) of any hoistway, or inside floor openings not within 10 ft (3.05 m) of an exterior wall that does not extend beyond the top of the material stored?
[29 CFR 1926.250(b)(1)] ☐ ☐ ☐

8. Are noncompatible materials segregated in storage?
[29 CFR 1926.2509(b)(3)] ☐ ☐ ☐

9. Are bagged materials stacked by stepping back the layers and cross-keying the bags at least every 10 bags high?
[29 CFR 1926.250(b)(4)] ☐ ☐ ☐

10. Is the storage of material PROHIBITED on scaffolds or runways in excess of the supply needed for the immediate operation?
[29 CFR 1926.250(b)(5)] ☐ ☐ ☐

11. Are brick stacks limited to 7 ft (2.13 m) in height?
[29 CFR 1926.250(b)(6)] ☐ ☐ ☐

Note: When a loose brick stack reaches a height of 4 ft (1.22 m), it must be tapered back 2 in (5.08 cm) on every foot height above the 4-ft (1.22-m) level.

12. When masonry blocks are staked higher than 6 ft (2.83 m), is the stack tapered back one half block per tier above the 6-ft (2.83 m) level?
[29 CFR 1926.250(b)(7)] ☐ ☐ ☐

13. Are all nails withdrawn from lumber before stacking?
[CFR 1826.250(b)(8)(i)] ☐ ☐ ☐

14. Is lumber stacked on level and solidly supported sills?
[29 CFR 1926.250(b)(8)(ii)] ☐ ☐ ☐

15. Is lumber stacked in a stable, self-supporting manner?
[29 CFR 1926.250(b)(8)(iii)] ☐ ☐ ☐

16. Are all lumber piles 20 ft (6.1 m) or less in height?
[29 CFR 1926.250(b)(8)(iv)] ☐ ☐ ☐

17. Do lumber piles to be handled manually NOT exceed a stacked height of 16 ft (4.88 m)?
[29 CFR 1926.250(b)(8)(iv)] ☐ ☐ ☐

18. Is all structural steel, poles, pipe, bar stock, and other cylindrical material, unless racked, stacked, and blocked, stored so as to prevent spreading or tilting? [29 CFR 1926.250(b)(9)]   ☐ ☐ ☐

19. Are all masonry walls over 8 ft (2.43 m) in height braced to prevent overturning? [29 CFR 1926.706(b)]   ☐ ☐ ☐

Note: The American National Standards Institute's Standard for Concrete and Masonry Work, ANSI A10.9-1983 standard, recommends that "the support or bracing shall be designed by or under the supervision of a qualified person to withstand a minimum of 15 pounds per square foot. Local environmental conditions (e.g., strong winds) need to be considered in determining the bracing design. Braces or shores MUST be secured in position."

Source: Environmental & Occupational Health Science Institute, New Jersey Department of Education.

Standards (consensus)

ASME B56.1	Low Lift and High Lift Trucks
ASME B56.6	Rough Terrain Fork Lift Trucks
ANSI B56.2	Powered Industrial Trucks, Type Designation, and Area of Use
ANSI B56.3	Electric Battery-Powered Industrial Trucks
ANSI MH11.4	Forks and Forks Carriers for Powered Industrial Forklift Truck
ANSI/UL 558	Standard for Safety: Industrial Trucks, Internal Combustion Engine-Powered
NFPA No. 505-1969	Powered Industrial Trucks

4.3 Helicopters

Where it may be impractical or uneconomical to use a crane of any sort as a lifting device, it may prove necessary to take advantage of the versatility of a helicopter, which can reach points far higher than those attainable with the boom of a conventional crane or tower; and are also ideally suited for free egress into remote reaches.

Currently available commercial helicopter technology and advanced designs have combined to make the helicopter a tool that fulfills specific commercial needs at improved operating costs.

To use a helicopter successfully on a construction job requires a sound project organization, a proper operational plan, and strict control, plus enforcement of operating procedures and on-site safety measures. Failure in any of these areas will produce a breakdown in coordinated efforts, increasing the amount of flight and ground time. These, in turn, will escalate costs and possibly delay the project's scheduled completion.

The first thing a contractor must do is educate construction crews on the sequence of tasks and specific safety requirements before introducing the helicopter into operations. Make sure that every member of the construction team understands the importance of executing each step on time, so that there will be no delays to subsequent tasks.

More than with most equipment in the contractor's spread, project operations must be built around the helicopter, not the helicopter around the construction operations. Cost and availability make this essential.

Maintenance guidelines

Helicopter contractors normally operate their equipment for lifting on construction projects or in facility operations. Thus they are responsible for, and must provide, the proper maintenance inspection of the helicopters, prior to and during use.

INSPECTIONS	YES	NO	N/A
1. Is a briefing conducted prior to each day's operation—setting forth the plan of operation for the pilot and ground personnel? [29 CFR 1910.183(b)]	☐	☐	☐
2. Are all loads properly slung?	☐	☐	☐
▪ Are tag lines of a length that will not permit their being drawn up into the rotors?	☐	☐	☐
▪ Are pressed sleeve, swaged eyes, or equivalent means used for all freely suspended loads to prevent hand splices from spinning open or cable clamps from loosening? [29 CFR 1910.183(c)]	☐	☐	☐
3. Do all electrically operated cargo hooks have the electrical activating device so designed and installed as to prevent inadvertent operation?	☐	☐	☐
▪ Are these cargo hooks equipped with an emergency mechanical control for releasing the load?	☐	☐	☐
▪ Have the hooks been tested prior to each day's operation by a competent person to determine that the release functions properly, both electrically and mechanically? [29 CFR 1910.183(d)]	☐	☐	☐

4. Is personal protective equipment provided? ☐ ☐ ☐
   [29 CFR 1910.183(e)(1)]
5. Do employees, receiving the load, use the personal ☐ ☐ ☐
   protective equipment?
   - Does personal protective equipment consist of ☐ ☐ ☐
     complete eye protection and hardhats secured by
     chinstraps?
   - Are personnel PROHIBITED to wear loose-fitting ☐ ☐ ☐
     clothing, likely to flap in rotor downwash, and
     thus be snagged on the hoist line?
     [29 CFR 1910.183(e)(2)]
6. Are all necessary precautions taken to protect ☐ ☐ ☐
   employees from flying objects in the rotor downwash?
   - Is all loose gear within 100 ft (30.48 m) of the place ☐ ☐ ☐
     of lifting the load or depositing the load, or within
     all other areas susceptible to rotor downwash, shall
     be secured or removed?
     [29 CFR 1910.183(f)]
7. Is good housekeeping maintained in all helicopter ☐ ☐ ☐
   loading and unloading areas?
   [29 CFR 1910.183(g)]
8. Is the size and weight of loads checked before ☐ ☐ ☐
   connection to the helicopter?
   - Is the manner in which loads are connected to the ☐ ☐ ☐
     helicopter checked?
   - Is a lift PROHIBITED, if the helicopter operator ☐ ☐ ☐
     believes the lift cannot be made safely?
     [29 CFR 1910.183(h)]
9. Is a safe means of access provided for employees ☐ ☐ ☐
   working under hovering craft, hooking and unhooking
   loads, to reach the hoist line hook and engage or
   disengage cargo slings?
   - Are employees PROHIBITED to perform work ☐ ☐ ☐
     under hovering craft except when necessary to
     hook or unhook loads?
     [29 CFR 1910.183(i)]
10. Is static charge on the suspended load dissipated ☐ ☐ ☐
    with a grounding device before ground personnel
    touch the suspended load, unless protective rubber
    gloves are being worn by all ground personnel who
    may be required to touch the suspended load?
    [29 CFR 1910.183(j)]

11. Does the weight of an external load exceed the ☐ ☐ ☐
    helicopter manufacturer's rating?
    [29 CFR 1910.183(k)]

12. Are ground personnel PROHIBITED to attach hoist ☐ ☐ ☐
    wires or other gear to any fixed ground structure, or
    allowed to foul on any fixed structure—except for
    pulling lines or conductors that are allowed to
    "pay out" from a container or roll off a reel?
    [29 CFR 1910.183(l)]

13. Are ground personnel instructed, when visibility is ☐ ☐ ☐
    reduced by dust or other conditions, to exercise
    special caution to keep clear of main and stabilizing
    rotors?
    - Is action taken to ensure the ground crew take ☐ ☐ ☐
      special precaution to keep clear of main and
      stabilizing rotors?
    - Are measures taken to eliminate, as far as ☐ ☐ ☐
      practical, the dust or other conditions reducing
      the visibility?
      [29 CFR 1910.183(m)]

14. Are the air crew and ground personnel instructed on ☐ ☐ ☐
    the signal systems to be used?
    - Is the system reviewed with the employees in ☐ ☐ ☐
      advance of hoisting the load?
      (This applies to both radio and hand signal
      systems.)
    - Do hand signals, where used, follow OSHA ☐ ☐ ☐
      Helicopter Hand Signal 1910.1839n?
      [29 CFR 1910.183(n)]

15. Are employees PROHIBITED to approach within ☐ ☐ ☐
    50 ft (15.24 m) of the helicopter when the rotor blades
    are turning, unless the employee's work duties require
    presence in that area?
    [29 CFR 1910.183(o)]

16. Are employees instructed, and measures taken to ☐ ☐ ☐
    ensure, that whenever approaching or leaving a
    helicopter, which has its blades rotating, all employees
    shall remain in full view of the pilot and keep in a
    crouched position?

17. Are employees instructed, and measures taken to ensure, that no employee is permitted to work in the area from the cockpit or cabin rearward while blades are rotating, unless authorized by the helicopter operator to work there?
[29 CFR 1910.183(p)] ☐ ☐ ☐

18. Are sufficient ground personnel provided to ensure that helicopter loading and unloading operations can be performed safely?
[29 CFR 1910.183(q)] ☐ ☐ ☐

19. Is constant reliable communication maintained between the pilot and a designated employee of the ground crew who acts as a signalman during the period of loading and unloading? ☐ ☐ ☐
    - Is the signalman clearly distinguishable from other ground personnel?
    [CFR 1910.183(r)] ☐ ☐ ☐

20. Are open fires PROHIBITED in areas where they could be spread by the rotor downwash?
[29 CFR 1910.183(s)] ☐ ☐ ☐

INDUSTRY RECOMMENDATIONS   YES   NO   N/A

1. Are all aerial lift staging areas located outside of a 20-ft (6.01-m) safety zone around overhead power lines and are clearly marked? ☐ ☐ ☐
    - Is an extended safety zone used when aerial lifts cannot be done downwind of energized overhead power lines? ☐ ☐ ☐

2. Are workers provided with two-way radios for communicating with the helicopter pilot during aerial lifts? ☐ ☐ ☐

3. Is a spotter present to provide directions to the helicopter pilot during aerial lift operations? ☐ ☐ ☐
    - Is the spotter and pilot familiar with both verbal commands and hand signals? ☐ ☐ ☐

4. Do workers use insulated footwear and gloves while working near transmission lines? ☐ ☐ ☐

5. Do first aid plans minimize response times during emergencies? ☐ ☐ ☐

6. Are nonconductive cables used when airlifting loads near overhead power lines? ☐ ☐ ☐

Lifting Equipment

Source: *Helicopter Association International Safety Manual*, Helicopter Association International.

Helicopter hand signals

Regulations

29 CFR 1910
  Subpart N
   .183 Helicopters
   .184 Slings
29 CFR 1926
  Subpart N Cranes, Derricks, Hoists, Elevators, and Conveyors
   .551 Helicopters
  Subpart V Power Transmission and Distribution
   .958 External load helicopters (comply with provisions 1925.551 Subpart N)

Standards (consensus)

ASME-B 30.12	Handling Loads Suspended from Rotorcraft—using helicopters to lift loads
NFPA 407	Safeguards People and Property against the Fire Hazards of Aviation Fuels
SAE AIR 1076-1970	Aircraft Fire Protection, for Reciprocating and Gas Turbine Engine Installations (Reaffirmed: May 1989)

4.4 Jacks and Air Lifts

Jacks are designed and built for lifting a load for a short distance, such as placing cribbing, skids, and rollers —and they are also used for precise placement of heavy loads, such as beams, or for raising and lowering heavy loads a short distance. They are available with 5 tons to 100 tons capacities and include cable reel jacks, ratchet jacks, hydraulic jacks, and mechanical jacks, among others.

Although there are a number of different styles of jacks available, only heavy-duty hydraulic jacks or screw jacks should be used, with the number of jacks used determined by the weight of the load and the rated capacity of the jacks. MAKE SURE that every jack in use has a solid footing and is NOT susceptible to slipping.

Small-capacity jacks are normally operated through a rack bar or screw as the lifting mechanism, while large capacity jacks are usually operated hydraulically. Types of jacks typically used by steelworkers include

- Ratchet level jacks—Rack-bar systems having a rated capacity of 15 tons and having a foot lift close to the base of the jack that can be engaged, thus allowing low-clearance loads close to the base of the jack to be lifted.

- Push-and-pull jacks (also called Steamboat ratchets)—Ratchet-screw systems of 10-ton rated capacity with end fittings that permit pulling parts together or pushing them apart; primarily used for tightening lines or lashings and for spreading or bracing parts in bridge construction.

- Screw jacks—Units have a rated capacity of 12 tons. Approximately 13 in high when closed, having a safe rise of 7 in, these types are used for general purposes, including steel erection.

- Hydraulic jacks—Operate on the piston-cylinder principle, with oil pumped through a line into a liquid-tight cylinder that forces the piston to move against the load. (A slow leak ordinarily prevents the jacks from holding the load in an exact position of long periods.) Available in many different capacities (10–100 tons), hydraulic jacks are used for general purposes.

- Low-weight flat jacks—Range from 8 to 345 tons capacities. Their low height makes them capable of exerting extremely high trust forces, and they are normally used by the construction industry for a multitude of applications, such as lifting heavy structures, stressing operations, transferring existing loads to temporary or permanent supports and permanently preloading bridge bearings. When selecting a FLAT JACK, the ideal working range should be about 75 percent of the jacks' capacity.

Maintenance guidelines

Use a good checklist when performing preventive maintenance to make sure all the work items are performed, and make notes of any issues that should be checked into further.

- Keep the jack clean—Always clean the exterior of all hydraulic jacks after use.

Note: Keeping the exterior of the metal body clean not only removes debris, but also can identify any oil leaking from the cylinder.

- Store the jack in a dry area—Moist or wet areas can start to rust the body of the jack.

- Keep the ram retracted when not in use—Most steel rams are bare machined metal, with unprotected surface prone to rust and corrosion if left to the outside elements.

- Clean any rust that begins to form—Gently work the surface with a piece of emery cloth and apply high-grade machine oil to the metal surface with a clean rag.

- Use only approved hydraulic oil—Consult the manufacturer's recommendations when adding hydraulic oil.

Note: DO NOT use brake fluid—it contains alcohol that will quickly ruin the internal seals.

- Locate the oil-filling plug—On the side of the cylinder round body.
- DO NOT open the other screw plugs—Located on the square metal base of the jack. These ports contain the check valves for raising and lowering the jack. If the small internal springs and small metal balls (they hold in place) are lost due to removal, the jack will not operate.

Note: Keep the jack clean to be able to read the "fill" or "oil" label printed on the side of the jack cylinder.

- DO NOT lift too heavy items—Exceeding the jack's lift capacity places excessive pressure on internal seals and valves; it not only will damage the seals, but also may result in injury to the operator from the jack's failure.

Warning: Most damage occurs to a hydraulic jack by either removing the check valve plug or lifting too heavy items.

Regulations

This checklist covers some of the regulations for jacks, overhead hoists, and monorail hoists issued by the U.S. Department of Labor—OSHA under General Industry standards 29 CFR 1910.244 and 1910.308 and the Construction standards 29 CFR 1926.305 and 1926.554.

	YES	NO	N/A
1. Do jacks in use have a rating sufficient to lift and sustain the loads? [29 CFR 1910.244(a)(1)(i)]	☐	☐	☐
2. Is the rated load legibly and permanently marked in a prominent location on the jack by casting, stamping, or other suitable means? [29 CFR 1910.244(a)(1)(ii) and 1926.305(a)(1)]	☐	☐	☐
3. In the absence of a firm foundation, is the base of the jack blocked or cribbed? [29 CFR 1910.244(a)(2)(i) and 1926.305(c)]	☐	☐	☐

4. If the cap could slip, is a block placed between the cap and the load? [29 CFR 1910.244(a)(2)(i) and 1926.305(c)]	☐	☐	☐
5. Do all jacks have a positive stop to prevent overtravel? [29 CFR 1926.305(a)(2)]	☐	☐	☐
6. Are operators instructed to watch the stop indicator (which must be kept clean) in order to determine the limit of travel? [29 CFR 1910.244(a)(2)(ii)]	☐	☐	☐
7. After the load has been raised, is it required to be cribbed, blocked, or otherwise secured at once? [29 CFR 1910.244(a)(2)(iii) and 1926.305(d)(1)(i)]	☐	☐	☐
8. Are hydraulic jacks—exposed to freezing temperatures—supplied with adequate antifreeze liquid? [29 CFR 1910.244(a)(2)(iv) and 1926.305(d)(1)(ii)]	☐	☐	☐
9. Are all jacks properly lubricated at regular intervals? [29 CFR 1910.244(a)(2)(v) and 1926.305(d)(1)(iii)]	☐	☐	☐
10. Is each jack thoroughly inspected? [29 CFR 1910.244(a)(2)(vi) and 1926.305(d)(1)(iv)]	☐	☐	☐
11. Are jacks that are used constantly or intermittently at one locality thoroughly inspected at least every 6 months? [29 CFR 1910.244(a)(2)(vi)(a) and 1926.305(d)(1)(iv)(a)]	☐	☐	☐
12. Are jacks that are sent out thoroughly inspected when they are returned? [29 CFR 1910.244(a)(2)(vi)(b) and 1926.305(d)(1)(iv)(b)]	☐	☐	☐
13. Are jacks that are subjected to abnormal load or shock thoroughly inspected immediately before and immediately after use? [29 CFR 1910.244(a)(2)(vi)(c) and 1926.305(d)(1)(iv)(c)]	☐	☐	☐
14. Are repair or replacement parts examined for defects? [29 CFR 1910.244(a)(2)(vii) and 1926.305(d)(1)(v)]	☐	☐	☐
15. Are repairs made on disabled jacks before they are used again? [29 CFR 1910.244(a)(2)(viii) and 1926.305(d)(1)(vi)]	☐	☐	☐

LIFT-SLAB OPERATIONS-REGULATIONS YES NO N/A

1. Are lift-slab operations designed and planned by a registered professional engineer who has experience in lift-slab construction?
   [29 CFR 1926.705(a)] ☐ ☐ ☐

Note: Such plans and designs shall be implemented by the employer and shall include detailed instructions and sketches indicating the prescribed method of erection. These plans and designs shall also include provisions for ensuring lateral stability of the building/structure during construction.

2. Are jacks/lifting units marked to indicate their rated capacity as established by the manufacturer?
   [29 CFR 1926.705(b)] ☐ ☐ ☐

3. Are jacks/lifting units NOT loaded beyond their rated capacity as established by the manufacturer?
   [29 CFR 1926.705(c)] ☐ ☐ ☐

4. Does jacking equipment have the capacity to support at least two and one-half times the load being lifted during jacking operations?
   [29 CFR 1926.705(d)] ☐ ☐ ☐

5. Is jacking equipment NEVER overloaded?
   [29 CFR 1926.705(d)] ☐ ☐ ☐

Note: For the purpose of this provision, jacking equipment includes any load bearing component which is used to carry out the lifting operation(s)—including, but is not limited to, threaded rods, lifting attachments, lifting nuts, hook-up collars, T-caps, shear heads, columns, and footings.

6. Are jacks/lifting units designed and installed so that they will neither lift nor continue to lift when they are loaded in excess of their rated capacity?
   [29 CFR 1926.705(e)] ☐ ☐ ☐

7. Do jacks/lifting units have a safety device installed that will cause the jacks/lifting units to support the load in any position in the event any jack lifting unit malfunctions or loses its lifting ability?
   [29 CFR 1926.705(f)] ☐ ☐ ☐

Lifting Equipment

8. Are all jacking operations synchronized in such a manner to ensure even and uniform lifting of the slab?
[29 CFR 1926.705(g)]
☐ ☐ ☐

9. During lifting, are all points at which the slab is supported kept within 1/2 in of that needed to maintain the slab in a level position?
☐ ☐ ☐

10. Is a device installed, if leveling is automatically controlled, that will stop the operation when the 1/2 in tolerance set forth in paragraph (g) of this section is exceeded or where there is a malfunction in the jacking (lifting) system?
[29 CFR 1926.705(h)]
☐ ☐ ☐

11. Are manual controls, to maintain leveling, located in a central location and attended by a competent person while lifting is in progress?
[29 CFR 1926.705(i)]
☐ ☐ ☐

12. Is the maximum number of manually controlled jacks/lifting units on one slab limited to a number that will permit the operator to maintain the slab level within specified tolerances of paragraph (g) of this section, but in no case does that number exceed 14?
[29 CFR 1926.705(j)]
☐ ☐ ☐

13. Are NO employees, except those essential to the jacking operation, permitted in the building/structure while any jacking operation is taking place, unless the building/structure has been "reinforced sufficiently" to ensure its integrity during erection?
[29 CFR 1926.705(k)(1)]
☐ ☐ ☐

Note: The phrase "reinforced sufficiently to ensure its integrity" (in this paragraph) means that a registered professional engineer, independent of the engineer who designed and planned the lifting operation, has determined from the plans that if there is a loss of support at any jack location, that loss will be confined to that location and the structure as a whole will remain stable.

14. Are employees, NOT essential to the jacking operation, NEVER permitted immediately beneath a slab—under any circumstances—while it is being lifted?
[29 CFR 1926.705(k)(2)]
☐ ☐ ☐

Inspections

15. Does a jacking operation begin (for the purpose of paragraph [k] of this section) when a slab or group of slabs is lifted, and end when such slabs are secured, with either temporary or permanent connections?
[29 CFR 1926.705(k)(3)]
☐ ☐ ☐

Note: Employers, who comply with appendix A to 1926.705 shall be considered to be in compliance with the provisions of paragraphs (k)(1) through (k)(3) of this section.
[29 CFR 1926.705(k)(4)]

16. Are wedges secured by tack welding, or an equivalent method of securing, to prevent them from falling out of position when making temporary connections to support slabs?
[29 CFR 1926.705(l)]
☐ ☐ ☐

Note: Lifting rods MUST NOT be released until the wedges at that column have been secured.

17. Is all welding on temporary and permanent connections performed by a certified welder, familiar with the welding requirements specified in the plans and specifications for the lift-slab operation?
[29 CFR 1926.705(m)]
☐ ☐ ☐

18. Is the load transfer from jacks/lifting units to building columns NEVER executed until the welds on the column shear plates (weld blocks) are cooled to air temperature?
[29 CFR 1926.705(n)]
☐ ☐ ☐

19. Are jacks/lifting units positively secured to building columns so that they do not become dislodged or dislocated?
[29 CFR 1926.705(o)]
☐ ☐ ☐

20. Is equipment designed and installed so that the lifting rods cannot slip out of position or does the employer institute other measures, such as the use of locking or blocking devices that will provide positive connection between the lifting rods and attachments and will prevent components from disengaging during lifting operations?
[29 CFR 1926.705(p)]
☐ ☐ ☐

Lifting Equipment

Standards (consensus)

B30.1-2009 Jacks, Industrial Rollers, Air Casters, and Hydraulic Gantries

Includes provisions that apply to the construction, operation, inspection, testing, and maintenance of mechanical ratchet jacks, hand- or power-operated mechanical screw jacks, hand- or power-operated hydraulic jacks, air lifting bags, industrial rollers, air casters, and telescopic hydraulic gantry systems.

Chapter 5

Hoisting Equipment

5.1 Portable and Overhead Hoists

A handset to guide the hoist's carry bar into place eliminates the need to support any weight—thus reducing the pushing and pulling forces that can be involved in moving a mobile hoist into position—and making it less complicated to move the carry bar into the required position for hoisting.

In addition, since the track on which the hoist is mounted may be fixed to the ceiling, walls, or supported on uprights that are fixed to the floor, it is usually possible to install an overhead hoisting system in any type of building. The combined ease of transverse and the reduced manual handling required to maneuver a hoist into position greatly increases the versatility of an overhead hoist when compared to a mobile hoist.

Unfortunately, portable and overhead hoists of chain or wire rope type, operated manually or with electric or air power, often are some of the more neglected equipment in rigging operations—many times only getting attention after they fail or are no longer functioning.

Preventive maintenance and frequent inspection, however, can prevent costly downtime and potentially dangerous situations.

Lifting equipment to handle materials efficiently, and with minimum physical effort, is one of the most important considerations in industry today, for two reasons:

- Growing pressure on management from workers, their trade unions, and safety representatives, to provide lifting equipment to ease their daily tasks
- Increasing awareness of the greater efficiency and reduced costs that can be obtained by employing modern mechanical handling equipment of all capacities

A variety of hoists are available for different lifting jobs, having load capacities ranging from 0.25 to 50 tons—with lifting ranges anywhere from 5 ft (1.52 m) to 20 ft (6.1 m). These include

- Ratchet lever
- Hand chain
- Electric chain
- Electric wire
- Air chain

Maintenance guidelines

Use a good checklist when performing preventive maintenance, to make sure all the work items are performed—and make notes of any issues that should be checked into further.

- Establish a procedure for routine maintenance inspection, based on usages—to enhance safety.
- Ensure that manually operated lever hoists meet or exceed the requirements of ANSME/ANSI B30.21.
  - Mark the rated load on the block.
  - Ensure that the manufacturer has tested the hoist with a test load of at least 125 percent of rated load.
  - Identify hoist controls to indicate function or direction of motion.
  - Mark hoists with identification, as follows:
    - Name of manufacturer
    - Manufacturer's model or serial number
  - Affix to the hoist of load block, in a readable position, a label of labels displaying precautionary information concerning operating procedures.
- Provide guarding for load sprockets pockets or teeth (to allow engagement of the load chain) against jamming of the load chain with the hoist mechanism under normal operating conditions.
- Check hoist for
  - Signs of missing, bent, or broken components
  - Rusty parts
  - Proper function of controls; all travel direction motions agree with hand chain pull
  - Proper operation (before beginning any lifting job)

Note: Remove handle, unloader plate, and dogs.

Hoisting Equipment

- Use a solvent to thoroughly remove old grease and let dry thoroughly before continuing use.
- Look at the handle for
  - Bending, cracks, deformation, or signs of cheater use
  - Impact damage on "stop areas"
  - Wear or damage to the end that engages the ratchet teeth
- Check the fit between the "dog" and its pivot pin for freedom to move, but without excessive clearance.
- Check the "release key" for damage.
  - Torsion spring, worn or bent
- Check closely the outer handle rim, where the "wear ring" rides.
  - Nicks or bends that can gouge the ring.
- Check the housing assembly, especially the "holding dogs," for damage—just like damage to the handle.
  - Two extension springs for wear or opening of the hook ends
  - Unloader plate for wear such as rounded corners and notches in the "window"
  - "Stop area" of the housing casting for impact damaged teeth on the ratchet wheel, for signs of wear or rounding of the top corners or undercuts in the teeth
- Check the shaft (removed) for wear or chipping caused by debris carried into the housing by the chain.
- Check points:
  - Hooks for signs of damage, cracks, nicks, gouges, deformation of the throat, bending or opening, wear on saddle or load bearing point, and twist.
  - Fit of hook shank mounting hole for wear caused by side loading or opening.
  - Hook nuts and pins that prevent the nut from loosening.
  - Hook safety latches are in place and functioning properly.

Note: Reassemble hoist in reverse order of disassembly—after all worn or damaged parts have been identified and replaced.

- Make sure when actuating force is removed, that it will automatically stop and hold any test load up to 125 percent of the rated load.
- Inspect load and hand chains—for nicks, gouges, and any type of deformation or damage to the chain.
  - Check for lubrication of load chain.
  - Check for open links or open connection links in hand chain.

- Check load chain.
  - Ensure it is properly reeved and is NOT kinked or twisted, and that its parts are NOT twisted about each other.
  - Pitched so as to pass over all load sprockets without binding.
  - Proof tested by chain or hoist manufacturer with load test of 150 percent of the rated load divided by the number of chain parts supporting the load.
  - Means to guard against load chain jamming in the load block, under normal operating conditions.
- Check hoist motion to ensure it does NOT have excessive drift and that stopping distance is normal.
- Check for any sign of oil or grease leakage on the hoist and hoist mechanism while operating the hoist.
- Check for any unusual sounds from the hoist and hoist mechanism while operating the hoist.
- Check that WARNING and other SAFETY labels are NOT missing and that they are legible.

Electric and air-operated hoists

- Follow the same steps as cleaning the chain hoist.
- Check wire rope (if applicable)—for broken wires, broken strands, kinks, and any deformation or damage to rope structure.
- Check limit devices.
  - Primary upper-limit device stops lifting motion of the hoist load block at the upper limit of travel.
  - Lower-limit device (if present) stops lowering motion of hoist load block at the lower limit of travel.

Note: On wire rope hoists, two wraps of the wire rope must remain at each anchorage on the drum; however, only one wrap of wire rope at each anchorage on the hoist drum is permitted if a lower-limit device is present.

- Make sure load blocks are of the enclosed type.
- Make sure means are provided to guard against wire rope (or load chain) jamming in the load block, under normal circumstances.

Electric-powered hoist

- Make sure braking system—when hoist is under normal operating conditions with rated load and test conditions, and test loads up to 125 percent of rated load—performs the following functions:

- Stops and holds the load when controls are released
- Limits the speed of load during lowering, with or without power, to a maximum speed of 120 percent of rated lowering speed for the load being handled
- Stops and holds the load hook in the event of a complete power failure
- Has thermal capacity for the frequency of operation required by the service

• Ensure braking system has provisions for adjustments where necessary to compensate for wear

Air-powered hoist

- Make sure braking system—when hoist is under normal cooperating conditions with rated load and test conditions, and test loads up to 125 percent of rate loads—performs the following functions:
  - Stops and holds the load hook when controls are released
  - Prevents an uncontrolled lowering of the load in the event of a loss of air pressure
  - Has thermal capacity for the frequency of operation required by the service
  - Has provision for adjustment where necessary to compensate for wear

Inspections

- The designated hoist operator is required to inspect overhead hoists before use, checking
  - Periodic inspection tag

Note: If it is missing or the inspection date has expired or is illegible, the equipment MUST be removed from service and the project hoisting and rigging inspector notified to schedule an inspection.

- Hoist main disconnect switch
- Pendant control or controllers or selector
- Wire rope or chain for damage—worn, cut, kinked, crushed, spooling, or birdcaging cable

Note: Documented 30-day inspection required.

- Hook for damage—bent, spread, cracked, and safety latch

Note: Documented 30-day inspection required.

- Upper-limit switch—hook block stop
- Reverse reeving—hoist cable direction

- Brake system—trolley, bridge, and hoist
- Trolley and bridge travel—MAKE SURE stops are in place and limit working, and travel path is clear of obstructions
- Hoist gearing system—any unusual noises
- Rails during operation—unusual wear or noise
- Lubrication—leaks, excess grease
- Impact rigging equipment to be used—slings, shackles, guide ropes, personnel protection equipment

- MAKE SURE to review weight limits.
- Inspection and testing requirements, specifically for overhead hoists, are NOT found in an OSHA standard. Some relevant information, however, can be found in the General Industry Standard 29 CFR 1910.179—Overhead and Gantry Cranes, while other general requirements can be found in the Construction Standard 29 CFR 1926.554.
- The standard that most specifically addresses the requirements of overhead hoists, however, is the Consensus Standard ASME/ANSI B30.16—Overhead Hoists (Underhung), a part of the ASME/ANSI B30 series on cableways, cranes, derricks, hoists, hooks, jacks, and slings.

Inspection schedules. Prior to the initial use and on regular intervals, hoists MUST be inspected by a designated person to verify compliance with ASME/ANSI B30.16, which contains the specific inspection requirements classified into FREQUENT inspections that do not require documentation and PERIODIC inspections that require documentation. The interval between inspections depends on the service of the hoist.

Check the owner's manual, specific to the hoist for inspection and maintenance requirements that should be based on the B30.16 standard requirements.

HAND CHAIN-OPERATED HOISTS   YES   NO   N/A

1. Does inspection and testing of hand chain-operated hoists, before use, include
    - All functional operating mechanisms for maladjustment and unusual sounds?   ☐ ☐ ☐
    - Hoist braking system for proper operation?   ☐ ☐ ☐
    - Hooks in accordance with ASME B30.10?   ☐ ☐ ☐
    - Hook latch operation, if used?   ☐ ☐ ☐
    - Load chain in accordance with paragraph B 30.16-2.5.1 or 16-2.6.1?   ☐ ☐ ☐
    - Load chain reeving for compliance with hoist manufacturer's recommendations?   ☐ ☐ ☐
    - Periodic inspection (refer to paragraph B30.16-2.1.3)?   ☐ ☐ ☐

2. Do frequent inspections and testing of hand chain-operated hoists include checking for
   - Evidence of loose bolts, nuts, or rivets? ☐ ☐ ☐
   - Evidence of worn, corroded, cracked, or distorted parts such as load blocks, suspension housing, hand chain wheels, chain attachments, clevises, yokes, suspension bolts, shafts, gears, bearings, pins, rollers, and locking and clamping devices? ☐ ☐ ☐
   - Evidence of damage to hook-retaining nuts or collars and pins, and welds or rivets used to secure the retaining numbers? ☐ ☐ ☐
   - Evidence of damage to or excessive wear of load sprockets, idler sprockets, or hand chain wheel? ☐ ☐ ☐
   - Evidence of worn, glazed, or oil-contaminated friction discs; worn pawls, cams, or ratchet; and corroded, stretched, or broken pawl springs in brake mechanism? ☐ ☐ ☐
   - Evidence of damage to supporting structure or trolley, if used? ☐ ☐ ☐
   - Label or labels required by paragraph B30.16-1.1.4 for legibility? ☐ ☐ ☐
   - End connections of load chain? ☐ ☐ ☐

Electric or Air-Powered Hoists

3. Does inspection and testing of electric or air hoists, before use, include
   - All functional operating mechanisms for maladjustment and unusual sounds? ☐ ☐ ☐
   - Limit devices for operation? ☐ ☐ ☐
   - Hoist braking system for proper operation? ☐ ☐ ☐
   - Air lines, valves, and other parts for leakage? ☐ ☐ ☐
   - Hooks in accordance with ASME B30.10? ☐ ☐ ☐
   - Hook latch operation, if used? ☐ ☐ ☐
   - Hook rope in accordance with paragraph B30.16-2.4.1(a)? ☐ ☐ ☐
   - Load chain in accordance with paragraph B30.16-2.5.1 or 16-2.6.1? ☐ ☐ ☐
   - Rope or load chain reeving for compliance with hoist manufacturer's recommendations? ☐ ☐ ☐
   - Periodic inspection (Refer to paragraph B30.16-2.1.3)? ☐ ☐ ☐

4. Do frequent inspections of electric or air-powered hoists include checking for
   - Hoist rope in accordance with paragraph B30-16-2.4.1(a)? ☐ ☐ ☐
   - Evidence of loose bolts, nuts, or rivets? ☐ ☐ ☐
   - Evidence of worn, corroded, cracked, or distorted parts such as load blocks, suspension housing, chain attachments, clevises, yokes, suspension bolts, shafts, gears, bearings, pins, rollers, and locking and clamping devices? ☐ ☐ ☐
   - Evidence of damage to hook-retaining nuts or collars and pins, and welds or rivets used to secure the retaining numbers? ☐ ☐ ☐
   - Evidence of damage to or excessive wear of load sprockets, idler sprockets, and drums or sheaves? ☐ ☐ ☐
   - Evidence of excessive wear on motor or load brake? ☐ ☐ ☐
   - Electric apparatus for signs of pitting or any deterioration of visible controller contacts? ☐ ☐ ☐
   - Evidence of damage to supporting structure or trolley, if used? ☐ ☐ ☐
   - Function labels on pendant control stations for legibility? ☐ ☐ ☐
   - Label or labels required by paragraph B30.16-1.1.4 for legibility? ☐ ☐ ☐
   - End connections of rope or load chain? ☐ ☐ ☐

Source: American Society of Mechanical Engineers (ASME).

Visual inspection by a designated person, recording existing conditions, provides the basis for a continuing evaluation.

REGULATIONS   YES   NO   N/A

Overhead Hoists

1. Is each overhead electric hoist equipped with a limit device to stop the hook at its highest and lowest point of safe travel?   ☐ ☐ ☐

2. Will each hoist automatically stop and hold any load up to 125 percent of its rated load if its actuating force is removed?   ☐ ☐ ☐

3. Is the rated load of each hoist legibly marked and visible to the operator?   ☐ ☐ ☐

4. Are stops provided at the safe limits of travel for trolley hoists? ☐ ☐ ☐

5. Are the controls of hoists plainly marked to indicate the direction of travel or motion? ☐ ☐ ☐

6. Is each cage-controlled hoist equipped with an effective warning device? ☐ ☐ ☐

7. Are close-fitting guards or other suitable devices installed on each hoist to ensure that hoist ropes will be maintained in the sheave grooves? ☐ ☐ ☐

8. Are all hoist chains or ropes long enough to handle the full range of movement of the application while maintaining two full wraps around the drum at all times? ☐ ☐ ☐

9. Are guards provided for nip points or contact points between hoist ropes and sheaves permanently located within 7 ft (2.13 m) of the floor, ground, or working platform? ☐ ☐ ☐

10. Are workers PROHIBITED from using chains or rope slings that are kinked or twisted or the hoist rope or chain wrapped around the load as a substitute for a sling? ☐ ☐ ☐

11. Is the operator instructed to avoid carrying loads above people? ☐ ☐ ☐

12. Are only employees who have been trained in the proper use of hoists allowed to operate them? ☐ ☐ ☐

Electric Monorail Hoists

13. Is a readily accessible disconnecting means provided on an electric monorail hoist between the runway contact conductors and the power supply?
[29 CFR 1910.306(b)(1)(i)] ☐ ☐ ☐

14. Is another disconnecting means (capable of being locked in open position) provided on an electric monorail hoist in the lead from runway contact conductors or other power supply?
[29 CFR 1910.306(b)(1)(ii)] ☐ ☐ ☐

Note: See 29 CFR 1910.306(b)(1)(ii)(B)(1 to 3) for exceptions.

15. Is there a means provided at the operating station to open the power circuit to all motors of the crane or monorail hoist—if the additional disconnecting means, required in question 2, is NOT readily accessible from the crane or monorail hoist operating station?
[29 CFR 1910.306(b)(1)(ii)(A)] ☐ ☐ ☐

16. Is a minimum of 2 ft 6 in (76.2 cm) of working space provided on an electric hoist, in the direction of access to live parts for examination, adjustment, servicing, or maintenance?
[29 CFR 1910.306(b)(3)] ☐ ☐ ☐

17. Are door(s) removable or capable of opening at least 90 degrees on an electric hoist where controls are enclosed in cabinets?
[29 CFR 1910.306(b)(3)] ☐ ☐ ☐

18. Is a limit switch or other device provided on an electric hoist, to prevent the load block from passing the safe upper limit of travel of any hoisting mechanism?
[29 CFR 1910.306(b)(2)] ☐ ☐ ☐

19. Is the safe working load of an overhead hoist, as determined by the manufacturer, indicated on the hoist? ☐ ☐ ☐
    - Is the weight of the load ALWAYS maintained at or below the safe working load?
    [29 CFR 1926.554(a)(1)] ☐ ☐ ☐

20. Does the supporting structure, to which an overhead hoist is attached, have a safe working load at least equal to that of the hoist?
[29 CFR 1926.554(a)(2)] ☐ ☐ ☐

21. Does the support for an overhead hoist allow free movement of the hoist so that the hoist can line itself up with the load?
[29 CFR 1926.554(a)(3)] ☐ ☐ ☐

22. Is the overhead hoist installed only in locations that will permit the operator to stand clear of the load at all times?
[29 CFR 1926.554(a)(4)] ☐ ☐ ☐

23. Do overhead hoists meet the applicable requirements for construction, design, installation, testing, inspection, maintenance, and operation as prescribed by the manufacturer?
[29 CFR 1926.554(a)(6)] ☐ ☐ ☐

24. Are overhead hoist hooks equipped with a hook latch safety device? ☐ ☐ ☐

Note: OSHA 1910.181(j)(2)(ii) only specifies "Safety latch type hooks shall be used wherever possible."

Air-Powered Hoists

25. Are air hoists connected to an air supply of sufficient capacity and pressure to operate the hoist safely?
[29 CFR 1926.554(a)(5)] ☐ ☐ ☐

26. Are air hoses supplying air to air hoists connected by a positive method to prevent their becoming disconnected during use?
[29 CFR 1926.554(a)(5)] ☐ ☐ ☐

Standards (consensus)

ASME B30.16 (R2007)	Overhead Hoists (Underhung)
	Applies to the construction, installation, operation, inspection, testing, and maintenance of hand chain-operated chain hoists and electric and air-powered chain and wire rope hoists used for, but not limited to, vertical lifting and lowering of freely suspended, unguided loads which consist of equipment and materials.
B30.21-1982	Manually Lever-Operated Hoists
B30.7	Base-Mounted Drum Hoists (Drum hoists mounted on a foundation)
ASME/ANSI	
B30.16-1987	Overhead Hoists (Underhung)
B 29.24	Overhead Hoists, Roller Load Chains (Configurations of roller load chains)

ANSI/ASME

HST-3M-1985	Manually Lever-Operated Chain Hoists (Performance Standard)
HST-2M-1993	Hand Chain Manually Operated Chain Hoists (Performance Standard)
HST-1M-1982	Electric Chain Hoist (Performance Standard)
HST-4M-1985	Overhead Electric Wire Rope Hoists (Performance Standard)
HST-5M-1985	Air-Powered Chain Hoists (Performance Standard)
HST-6M-1986	Air-Powered Wire Rope Hoists (Performance Standard)

5.2 Material and Personnel Hoists

Damaged or faultily operated hoists represent a danger not only to the hoist's operator, but also to the employees who may be unlucky enough to be standing nearby when it drops its load.

To ensure the safe operation of new hoists, on installation completion and before being placed in service, material and personnel hoists MUST be subjected to an acceptance inspection and test in the field to determine that all parts of the installation conform to the applicable requirements of OSHA standards and that all safety equipment functions as required.

A designated, qualified person MUST make a similar acceptance inspection and test following a major alteration of an existing installation and at any time the hoist is moved to a new location. A jump of the tower after initial installation, however, shall NOT be considered a major alteration.

Records MUST be maintained and kept on file at the job site for the duration of the job.

The employer MUST comply with the manufacturer's specifications and limitations applicable to the operation of all hoists and elevators. Where manufacturer's specifications are not available, the limitations assigned to the equipment shall be based on the determinations of a professional engineer competent in the field.

Maintenance guidelines

Use a good checklist when performing preventive maintenance, to make sure all the work items are performed—and make notes of any issues that should be checked into further.

- Identify the year of manufacture of the hoist.
- Comply with manufacturer's specifications and limitations applicable to the operation of all hoists.

Note: Where manufacturer's specifications are not available, the limitations assigned to the equipment MUST be based on the determinations of a professional engineer competent in the field.

- Check that the hoist has been maintained in accordance with the manufacturer's recommendations and ANSI/ASME B30.21-2005.
- Ensure a designated, qualified person performs periodic inspections and documents the results.
- Inspect the hoist at least every 6 months and log the results.
- DO NOT lubricate any part of the hoist except the wire rope and the snap hook.

Note: Hoist-winding mechanism is virtually maintenance free, with all bearings lubricated for life.

- Mark appropriate month and year on the inspection sheet and permanently on the hoist—using a steel stamp, taking care to not damage the hoist housing.
- Ensure that ongoing short-term maintenance requirements, including hoist operator's daily checks, have been carried out, following manufacturer's inspection instructions on the specification and instruction labels.
- Examine hoists for excessive wear, damage, alteration, or missing parts.

Note: Routine inspection must be performed at regular intervals and before and after lengthy periods of storage. Frequency of periodic inspection depends on the severity of environmental conditions and frequency of use of the unit, but must be performed at least every 6 months.

- Identify items that need to be maintained after every 40 hours of operation or in no case at a greater time interval than monthly.
- Verify that major inspection requirements have been identified and carried out, including annual inspection requirements.

116 Inspections

- Ensure that major strip down and inspections are performed after the first 10 years of service and again every 5 years thereafter.
- DO NOT operate a hoist where these requirements have not been met—until the required level of maintenance is complete.
- Ensure that maintenance records are available on site for inspection by safety field officers or safety representatives (or both).
- Ensure that requirements for maintenance and maintenance records are met for any hoist under control of hoist-hire company and hoist owners—prior to further supply for use.
- Ensure that hoist operators carry out the daily checks before a hoist is used and at the beginning of each shift.
- MAKE SURE that all safety-related problems, identified in such checks, are rectified prior to using the hoist.
- Notify the principal contractor promptly of any matter identified by the hoist operator indicating a need for repairs or maintenance.

Regulations

Below are listed just a few points to consider when putting together or reviewing the efficacy of a facility's hoist and auxiliary equipment safety program:

MATERIAL HOISTS	YES	NO	N/A
1. Are workers prohibited from riding the hoist?	☐	☐	☐
2. Is overhead protection provided over the case or platform and the operator's position?	☐	☐	☐
3. Is a load-rating plate attached to the hoist?	☐	☐	☐
4. Has wire rope been inspected for harmful defects?	☐	☐	☐
5. Are there at least three full wraps on the winding drum when the platform is at the lowest point of travel?	☐	☐	☐
6. Is there at least 3 ft of clearance between the cathead sheave and the top of cage when it is at the uppermost terminal or landing?	☐	☐	☐
7. Are sheave bearings well lubricated?	☐	☐	☐
8. Are brakes capable of stopping and holding 125 percent of the rated load?	☐	☐	☐
9. Does the operator remain at the controls while the load is suspended or the master clutch is engaged?	☐	☐	☐
10. Are gears on the hoisting machine well guarded?	☐	☐	☐

11. Are entrances to hoist way guarded or barricaded? ☐ ☐ ☐
12. Are nip points or contact points—between hoist ropes and sheaves that are permanently located within 7 ft (2.13 m) of the floor, ground, or working platform—guarded? ☐ ☐ ☐
13. Are riggers PROHIBITED
    - To use chains or rope slings that are kinked or twisted? ☐ ☐ ☐
    - To use the hoist rope or chain wrapped around the load as a substitute, for a sling? ☐ ☐ ☐
14. Is the operator instructed to avoid carrying loads over people? ☐ ☐ ☐
15. Are all hoist towers, masts, guys or braces, counterweights, drive machinery supports, sheave supports, platforms, supporting structures, and accessories designed by a licensed engineer? ☐ ☐ ☐
16. Is a copy of the hoist-operating manual available? ☐ ☐ ☐
17. Is a periodic inspection of material hoists conducted prior to initial use and at least monthly thereafter, by a qualified person? ☐ ☐ ☐
    - Are periodic inspections conducted according to the manufacturer's specifications? ☐ ☐ ☐
    - At a minimum, do inspections cover sheaves, racks, pinions, guy ties, connections, miscellaneous clamps, braces, and similar parts? ☐ ☐ ☐
18. Are preoperational inspections conducted prior to every operation (shift) of the hoist? ☐ ☐ ☐
19. Do all floors and platforms have slip-resistant surfaces? ☐ ☐ ☐
20. Are landings and runways adequately barricaded and is overhead protection provided where needed? ☐ ☐ ☐
21. Are hoisting ropes installed according to the manufacturer's instructions? ☐ ☐ ☐
22. Are there at least three full wraps of cable on the drums of the hoist at all times? ☐ ☐ ☐
23. Does the rope or crane manufacturer specify the drum end of the rope anchored to the drum by an arrangement? ☐ ☐ ☐
24. Are personnel PROHIBITED from riding on material hoists or other hoisting equipment? ☐ ☐ ☐

25. Are operating rules posted at the hoist operator's station? ☐ ☐ ☐
26. Are air-powered hoists connected to an air supply of sufficient capacity and pressure to safely operate the hoist? ☐ ☐ ☐
27. Are pneumatic hoses secured by some positive means to prevent accidental disconnection? ☐ ☐ ☐

Personnel hoists. To ensure the safe operation of personnel hoists acceptance Inspections and Tests MUST be made by a qualified, designated person. Records MUST be maintained and kept on file at the job site for the duration of the job.

	YES	NO	N/A
1. Are all parts of a personnel hoist installation inspected for conformity with the applicable requirements of the manufacturer?	☐	☐	☐
2. Are safety tests performed for			
• Car—with rated load in the car?	☐	☐	☐
• Oil buffers under the car—with a rated load on the car traveling at rated speed?	☐	☐	☐
• Counterweight, where provided—with no load in the car?	☐	☐	☐
• Oil buffers under the counterweight—with no load on the car and the counterweight traveling at rated speed?	☐	☐	☐
• Governor-tripping speed measurement?	☐	☐	☐
• Governor over speed switch operation?	☐	☐	☐
• Safety device or equipment inoperative or which require removal or resetting of devices or equipment?	☐	☐	☐

Standards (consensus)

ASME B30.16-2003   Overhead Hoists (Underhung)

Applies to the construction, installation, operation, inspection, and maintenance of hand chain-operated chain hoists, and electric- and air-powered chain and wire rope hoists used for, but not limited to, vertical lifting and lowering of freely suspended, unguided loads, which consist of equipment and materials.

ASME B30.21-2005	Manually Lever-Operated Hoists
	Applies to the construction, installation, operation, inspection, and maintenance of ratchet and pawl and friction brake type manually lever-operated chain, wire rope, and web strap hoists used for lifting, pulling, and tensioning applications.
ASME B30.7-2001	Base-Mounted Drum Hoists
	Applies to base-mounted drum hoists arranged for mounting on a foundation or other supporting structure for lifting or lowering loads, to derrick swingers, and to any variations which retain the same fundamental characteristics.
ASME HST-1M-1999 (R2004)	Performance Requirements for Electric Chain Hoists
	Applies to vertical lifting service involving material handling of freely suspended (unguided) loads using load chain of the roller or welded link types with one of the following types of suspension: lug, hook or clevis, and trolley.
ASME HST-2M-1999 (R2004)	Performance Standard for Hand Chain Manually Operated Chain Hoists
	Applies to chain, manually operated chain hoists for vertical lifting service involving material handling of freely suspended (unguided) loads using welded link-type load chain as a lifting medium.
ASME HST-3M-1999 (R2004)	Performance Standard for Manually Lever-Operated Chain Hoists
	Requirements for manually lever-operated chain hoists, used for lifting-, pulling-, and tensioning-type applications, include ratchet and pawl operation with
	Roller-type load chain
	Welded link-type load chain

ASME HST-4M-1999 (R2004)	Performance Standard for Overhead Electric Wire Rope Hoists
	Applies to electric wire rope hoists for vertical lifting service involving material handling of freely suspended (unguided) loads using wire rope with one of the following types of suspension: lug, hook, trolley, or base or deck mounted, but DOES NOT include base-mounted drum hoists of the type covered by ASME B30.7 (5) wall or ceiling mounted.
ASME HST-5M-1999 (R2004)	Air Chain Hoists
	Requirements for air-powered chain hoists for vertical lifting service involving material handling of freely suspended (unguided) loads using load chain of the roller or welded link types with one of the following types of suspension: lug, hook or clevis, or trolley.
ASME HST-6M-1999 (R2004)	Performance Standard for Air Wire Rope Hoists
	Requirements for air wire rope hoists for vertical lifting service involving material handling of freely suspended (unguided) loads using wire rope as the lifting medium with one of the following types of suspensions: lug, hook or clevis, trolley, or base or deck mounted, but DOES NOT include base-mounted drum hoists of the type covered by ASME B30.7 (5) wall or ceiling mounted.
ANSI A10.22-1990 (R1998)	Safety Requirements for Rope-Guided and Non-Guided Worker's Hoists
	Minimum safety requirements for temporary personnel hoisting systems used for the transportation of persons to and from working elevations during normal construction and demolition operations, including maintenance, and is restricted to use in special situations.

ANSI A10.4-2004 Safety Requirements for Personnel Hoists and Employee Elevators

Applies to the design, construction, installation, operating, testing, maintenance, alterations, and repair of hoists and elevators that

Are NOT an integral part of buildings

Are installed inside or outside buildings or structures during construction, alteration, demolition, or operation

Are used to raise and lower workers and other personnel connected with or related to the structure

Note: These personnel hoists and employee elevators may also be used for transporting materials.

ANSI/ASSE A10.5-2006 Safety Requirements for Material Hoists

Applies to material hoists used to raise or lower materials during construction, alteration, or demolition.

Note: It is not applicable to the temporary use of permanently installed personnel elevators as material hoists.

UL 1323-2004 Scaffold Hoists

Requirements cover manually and power-operated-type portable hoists intended for use with scaffolds suspended by wire ropes. This standard covers electrically powered hoists rated 250 V or less to be used in non-hazardous environmental locations in accordance with the National Electrical Code, NFPA 70.

5.3 Winches and Drums

Although a winch is basically used for pulling a load, some winches have a "Lifting" and a "Pulling" capacity ratings. Such winch/hoist devices are made to the buyer's specs, when ordering them, and designed to do duel industrial jobs.

Winches

Each winch is supplied from the factory with the warning tag and label. If the tag or label is not attached to the unit, order a new tag or label and install it. Read and obey all warnings and other safety information attached to the winch to make an operator aware of dangerous practices to avoid, but NOT limited to the following selected list:

- Only allow persons instructed in safety and operation of the winch to operate and maintain the winch.
- Only operate a winch if you are physically fit to do so.
- When a DO NOT OPERATE sign is placed on the winch, DO NOT operate the winch until the sign has been removed by designated personnel.
- Before each shift, check the winch for wear and damage. Never use a winch that inspection indicates is worn or damaged.
- Never lift a load greater than the rated capacity of the winch. See warning labels and tags attached to winch.
- Keep hands, clothing, etc., clear of moving parts.
- Never place your hand in the throat area of a hook or near wire rope spooling onto or off of the winch drum.
- Always rig loads properly and carefully.
- Be certain the load is properly seated in the saddle of the hook. Do not tip-load the hook as this leads to spreading and eventual failure of the hook.
- DO NOT "side pull" or "yank."
- MAKE SURE everyone is clear of the load path. DO NOT lift a load over people.
- Never use the winch for lifting or lowering people, and never allow anyone to stand on a suspended load.
- Ease the slack out of the wire rope when taking up wire rope. Do not jerk the load.
- Do not swing a suspended load.
- Never suspend a load for an extended period of time.
- Never leave a suspended load unattended.
- Pay attention to the load at all times when operating the winch.
- After use, properly secure winch and all loads.

- The operator must maintain an unobstructed view of the load at all times.
- Never use the wire rope as a sling.

Source: American Society of Mechanical Engineers.

Drum systems

Drum crushing is a rope condition sometimes observed that indicates deterioration of the rope. Spooling is that characteristic of a rope affecting how it wraps onto and off a drum. Spooling is affected by the care and skill with which the first larger of wraps is applied on the drum. Manufacturer's criteria during inspection usually specify

- Minimum number of wraps to remain on the drum
- Condition of drum grooves
- Condition of flanges at the end of drum
- Rope end attachment
- Spooling characteristics of rope
- Rope condition

Maintenance guidelines

Use a good checklist when performing preventive maintenance, to make sure all the work items are performed—and make notes of any issues that should be checked into further.

A winch, like any other type of machinery, needs periodic maintenance and inspection to maintain its performance capabilities, give lasting value, and ensure safe mechanical workings.

- MAKE SURE that the winch and cable are maintained properly.
- Pay especially close attention to the clutch, making sure that it fully engages when shifted.
- Do not attempt to disengage the clutch when a load is on the winch.
- Avoid shock loads that impose a strain on the winch many times the actual weight of the load and can cause failure of the cable or of the winch.
- Inspect winches daily to ensure that
  - There are no oil leaks present.
  - All mounting bolts and other fasteners are tight.
  - Wire rope is in good condition.
- Periodically service the winch, including changing the oil in both the gearbox and the brake section.

Inspections

Note: Severity of use will determine the need for oil changes, but it should be checked at a minimum of every 500 operating hours and changed every 1000 hours of operation. Factors such as extremely dirty conditions or widely varying temperature changes may dictate even more frequent servicing.

- Ensure complete teardowns and component inspections.

Note: Severity and frequency of use will determine how often teardown should be done.

- If the equipment on which this winch is mounted is subject to inspection standards, those standards must also apply to the winch and be followed.
- If oil changes reveal significant metallic particles, a teardown and inspection must be made to determine the source of wear.
- Check gearbox oil levels weekly. If the oil level does not show a satisfactory amount, refill oil according to the individual winch manual.
- Check brake oil level and fill or replace if oil shows significant metallic particles.
- Lubricate all bushings equipped with grease zerks with good quality lithium-based chassis lube.
- Lubricate the cable, based on the wire rope supplier's recommendations, which operates in ambient temperatures from −10°F to +110°F.

Note: Contact manufacturer if winch is to be operated in temperatures outside this range.

- MAKE SURE the hydraulic system driving the winch uses only high-quality hydraulic oils from reputable suppliers that contain additives to prevent foaming and oxidation in the system.
- MAKE SURE all winch hydraulic systems are equipped with a return line filter capable of filtering 10-μm particles from the system.
- Keep at least five wraps of cable on the drum to ensure that the cable doesn't come loose.

Note: Cable anchors on winches are not designed to hold the rated load of the winch.

- Stay clear of suspended loads and of cable under tension.

Regulations

OSHA Regulation 29 CFR 1926.553 mandates guarding exposed moving parts of base-mounted drum hoists, such as gears, projecting screws, setscrews, chain, cables, chain sprockets, and reciprocating or rotating parts, which constitute a hazard.

In addition, all base-mounted drum hoists in use MUST meet the applicable requirements for design, construction, installation, testing, inspection, maintenance, and operations, as prescribed by the manufacturer.

WINCHES AND DRUMS	YES	NO	N/A
1. Are operators trained and qualified to work with winches and drums?	☐	☐	☐
2. Are operators PROHIBITED to winch for lifting, supporting, or transporting people or lifting loads over people?	☐	☐	☐
3. Are winches ALWAYS inspected and maintained in accordance with ANSI B30.7 Safety Code and any other applicable safety codes and regulations?	☐	☐	☐
4. Do the supporting structures and load-attaching devices, used in conjunction with the winch, provide an adequate safety factor to handle the rated load, plus the weight of the winch and attached equipment?	☐	☐	☐
5. Are hazardous moving parts (gears, drums, fly wheels, piston arms, moving cables, spokes, knobs, projections on hand wheels, etc.) adequately guarded?	☐	☐	☐
6. Is the drum end of the rope anchored to the drum by an arrangement specified by the crane or rope manufacturer?	☐	☐	☐
7. Are at least three full turns of cable retained on drums at all times?	☐	☐	☐
8. Is an inspection made of rigging equipment by a competent person before use on each shift?	☐	☐	☐
9. Do all slings (welded alloy steel chain, wire rope, and synthetic web) contain the required identification labels?	☐	☐	☐

10. When wedge socket-type fastening is used on wire rope, is the short end looped back (or have a separate piece of equal-size wire rope attached) and clipped with a U-bolt? ☐ ☐ ☐

11. Are drums, sheaves, and pulleys smooth and free of defects that may damage rigging? ☐ ☐ ☐

12. Is the winch shut down—power off—for servicing, or fueling, or both? ☐ ☐ ☐

13. On gasoline engine-powered winches, are the carburetors equipped with a back-fire trap flame arrestor? ☐ ☐ ☐

14. Is a fuel shut-off valve located at both the tank and engine? ☐ ☐ ☐

15. Are fuel tanks placed so that overflow or spills will not fall into/on electric equipment or the engines? ☐ ☐ ☐

16. On floating plant gasoline-powered winches, do carburetors have drip pans fitted with flame screens continuously emptied by suction from the intake manifold or a waste tank? ☐ ☐ ☐

Note: Not required on down draft carburetors.

17. When cranks are used on hand-powered winches or hoists, are the winces or hoists provided with positive self-locking dogs? ☐ ☐ ☐

18. Is a ratchet and pawl provided for releasing the load from the hoisting machinery brake? ☐ ☐ ☐

19. Is an inspection made of the cables at scheduled intervals? ☐ ☐ ☐

20. Are cracked spokes, hubs, or flanges in use? ☐ ☐ ☐

21. Is housekeeping maintained? ☐ ☐ ☐

22. Is a signal person provided when necessary? ☐ ☐ ☐

23. Is an operating procedure/manual available for the operator's use? ☐ ☐ ☐

24. Are test and inspection records maintained and available for review? ☐ ☐ ☐

25. Is machinery and equipment operated in a manner to ensure persons or property are not endangered? ☐ ☐ ☐

Recommended hand signals.

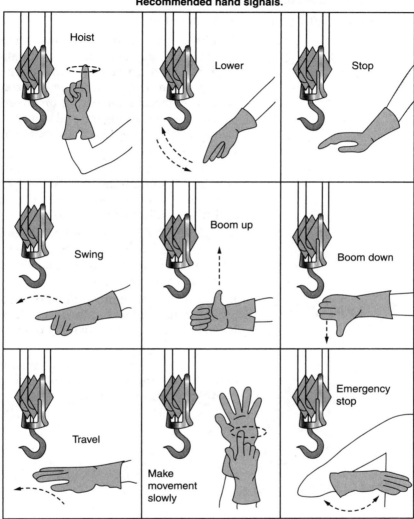

Hand Signals for Hoisting Operations

26. Are confined spaces labeled "Danger-Confined Space"? ☐ ☐ ☐
27. Are working/walking surfaces nonslip? ☐ ☐ ☐

Caution: If any malfunctions are observed or unusual noises are heard, STOP using the machinery and contact supervisor or safety coordinator immediately.

Source: USACE Safety and Health Requirements Manual (EM385-1-1).

Standards (consensus)

ANSI/ASME B30.7-2006   Base-Mounted Drum Hoists

Applies to the construction, installation, operation, inspection, and maintenance of base-mounted drum hoists arranged for mounting on a foundation or other supporting structure for lifting or lowering loads, to derrick swingers, and to any variations which retain the same fundamental characteristics.

The standard covers hoists that are powered by internal combustion engines, electric motors, compressed air, or hydraulics, and which utilized drums and rope.

It DOES NOT apply to overhead hoists, car pullers, barge pullers, truck body hoists, or other hoists or winches used exclusively in horizontal pulling applications.

It DOES NOT encompass all of the safety precautions and safeguards applicable when hoist loads consist wholly, or in part, of personnel.

It MAY NOT apply when base-mounted drum hoists are used as an integral part of other lifting equipment.

Chapter 6

Rigging Systems

6.1 Wire Rope and Wire Rope Slings and Meshes

Neglect and abuse are the two chief enemies of wire rope life. Abuse takes many forms; improper reeling or unreeling, wrong size or worn sheaves, improper storage, and bad splicing are a few. One costly form of neglect, however, is lack of proper field lubrication.

Improper use of hoisting equipment, including slings, may result in overloading, excessive speed (e.g., taking up slack with a sudden jerk, shock loading), or sudden acceleration or deceleration of equipment.

There are generally six types of slings: chain, wire rope, metal mesh, natural fiber rope, synthetic fiber rope, or synthetic web. Generally, they are grouped as: wire rope and mesh, alloy steel chain, and fiber rope web.

Each type has its own particular advantages and disadvantages. Factors to consider when choosing the best sling for the job include size, weight, shape, temperature, and sensitivity of the material being moved, and the environmental conditions under which the sling will be used.

An important factor to consider when using slings is the angle effect. Sling capacity is reduced when slings are used at an angle (i.e., two slings, or one sling in a basket hitch, attached to only one crane hook). How much it is reduced, however, depends on the degree of the angle.

To determine whether a sling will be rated high enough based on the angle between the sling leg and the horizontal, simply multiply the sling's rating by the appropriate factor in ANSI B30.9-1971 to obtain the sling's reduced rating.

Before using wire rope and wire rope slings, the employer MUST ascertain and adhere to the manufacturer's recommended ratings and MUST have such ratings available for inspection. When the manufacturer is unable to supply such ratings, however, the employer MUST use the tables for wire rope and wire rope slings, see ANSI B30.9-1971—Safety Standard for Slings.

Maintenance guidelines

Use a good checklist when performing preventive maintenance to make sure all the work items are performed, and make notes of any issues that should be checked into further.

Slings, like any machine, require regular maintenance. Lack of maintenance will increase the likelihood of a failure, which can be catastrophic, resulting in property damage, serious injuries, or death. Wire rope failure is usually cumulative; thus, it is important to inspect for overstressing, corrosion, severe heat, and machinery operation

- Follow manufacturer's specific recommendations and requirements pertaining to the use, maintenance, and inspection of wire rope and wire mesh slings. Manufacturer's recommended maintenance guidelines and scheduled inspections of all rigging systems include
  - Pre-use inspection—Performed by the operator, who will be using the equipment, prior to use each day or shift but DOES NOT require inspection documentation
  - Annual inspection—Performed by a qualified inspector, who marks the equipment that meets inspection criteria with an inspection sticker indicating the date of next inspection
  - Periodic inspections—Designated person MUST inspect slings and rigging hardware (at least annually) in accordance with ASME B30.9 and B30.26—basing inspection frequency on

  Frequency of use

  Severity of service conditions

  Nature of lifts being made

  Experience gained on the service life of slings used in similar applications

Note: Requirements for pre-use and periodic inspection of slings are contained in ASME B30.9 and 29 CFR 1910.184.

Mandatory Inspections

	Inspections	Load testing	Color coding
Wire rope slings	Every 12 mo. Diameter >5/8 in (16 mm)	Every 3 mo Diameter >1/2 in (12 mm)	Every 4 mo
Chain slings	Every 12 mo	Every 3 mo	Every 4 mo
Web slings	Every 12 mo	Every 3 mo	Every 4 mo

Note: OSHA requirements call not only for regular inspection and maintenance of all rigging systems, but also for documentation of all services rendered on that equipment.

- Visually inspect all slings and rigging hardware, prior to every use by the operator.
- Record all sling information on inspection documents; however, documentation of pre-use inspections is not required.
- DO NOT use equipment with expired inspections.
- Mark each sling in accordance with ASME B30.9 to show
  - Name or trademark of manufacturer
  - Rated loads for the type of hitch used and the angle upon which it is based
  - Diameter or size of sling
- Remove from service any slings that DO NOT legibly display the manufacturer's load rating tag.
- DO NOT USE, for any reason, slings and rigging hardware that appear to be damaged.
- Destroy and discard all damaged slings and rigging hardware, as well as when the inspector questions its condition.
- Ensure sling identification is legible and shows the rated capacities for each type of hitch (vertical, basket, and choke), prior to each use.

Note: Refer to ASME B30.20, ASME B30.26, and ASME B30.9 for further information on use and maintenance.

- Keep records of all inspections and maintenance work on slings and hardware.
- Visually inspect all slings and rigging hardware prior to every use by the operator in accordance with 29 CFR 1910.184(d).
- Store slings and rigging hardware in an area where the items will not be subjected to mechanical damage, corrosive action, moisture, and extreme temperatures or kinking.

Note: Some slings, when stored in extreme temperatures, will experience reduced performance.

Regulations

This checklist covers regulations issued by the U.S. Department of Labor, Occupational Safety and Health Administration (OSHA) under the general industry standards 29 CFR 1910.184 and the construction standards 1926.251. It applies to slings used with other equipment to move material by lifting or hoisting.

The regulations cited apply only to private employers and their employees, unless adopted by a state agency and applied to other groups such as public employees.

Numerous tables are included as part of 29 CFR 1910.184 and 1926.251 that relate to sling configuration, sling construction, sling diameter, and maximum load capacity. These tables have not been included as part of this checklist. For additional information, consult the OSHA regulations.

WIRE ROPE SLINGS YES NO N/A

1. Are wire rope sling capacity tags missing or illegible? ☐ ☐ ☐

2. Are wire rope slings used with loads at or below their rated capacities? ☐ ☐ ☐
[29 CFR 1910.184(f)(1) and 1926.251(c)(1)]

Note: Consult the OSHA regulations or the manufacturer's requirements for permitted load ratings.

3. Are fiber core wire rope slings permanently removed from service if they are exposed to temperatures above 200°F? ☐ ☐ ☐
[29 CFR 1910.184(f)(3) and 1926.251(c)(14)]

4. Are recommendations of the sling manufacturer followed when non-fiber core wire rope slings are used at temperatures above 400°F or below −60°F? ☐ ☐ ☐
[29 CFR 1910.184(f)(3) and 1926.251(c)(14)]

5. Is welding of end attachments performed before assembly of the sling? ☐ ☐ ☐
[29 CFR 1910.184(f)(4)(i) and 1926.251(c)(15)(i)]

6. Are all welded end attachments proof-tested by the manufacturer or equivalent entity at twice their rated capacity before their initial use, and is a certificate of the proof test available? ☐ ☐ ☐
[29 CFR 1910.184(f)(4)(ii) and 1926.251(c)(15)(ii)]

7. Are wire rope slings immediately removed from service if any of the following conditions are present:
    - Ten randomly distributed broken wires in one rope lay, or five broken wires in one strand in one rope lay? ☐ ☐ ☐

Note: One or more randomly distributed broken wires should prompt the user to conduct a more detailed inspection.

- Wear or scraping of one-third the original diameter of outside individual wires? ☐ ☐ ☐
- Kinking, crushing, birdcaging, or other damage, resulting in distortion of the wire rope structure? ☐ ☐ ☐
- Evidence of heat damage, acid, or caustic burns? ☐ ☐ ☐
- End attachments that are cracked, deformed, or worn? ☐ ☐ ☐
- Hooks that have been opened more than 15 percent of the normal throat opening (measured at the narrowest point), or hooks twisted more than 10 degrees from the plane of the unbent hook? ☐ ☐ ☐
- Corrosion of the rope or end attachments? ☐ ☐ ☐
  [29 CFR 1910.184(f)(5) and 1926.251(c)(4)(iv)]

8. Are protruding ends of strands in splices covered or blunted? ☐ ☐ ☐
   [29 CFR 1926.251(c)(2)]

Metal Mesh Slings

9. Does each metal mesh sling have a permanently affixed durable marking that states the rated capacity for vertical basket hitch and choker hitch loading? ☐ ☐ ☐
   [29 CFR 1910.184(g)(1)]

10. Do handles have a rated capacity at least equal to metal fabric and show no deformation after proof testing? ☐ ☐ ☐
    [29 CFR 1910.184(g)(2)]

11. When fabric and handles are joined,
    - Is the rated capacity of the sling the same (i.e., not reduced)? ☐ ☐ ☐
    - Is the load evenly distributed across the width of the fabric? ☐ ☐ ☐
    - Is the fabric protected from sharp edges? ☐ ☐ ☐
      [29 CFR 1910.184(g)(3)]

12. Are slings free of coatings that diminish the rated capacity of the sling prohibited? ☐ ☐ ☐
    [29 CFR 1910.184(g)(4)]

13. Are all new and repaired metal mesh slings and handles proof-tested by the manufacturer or other competent person at a minimum of one and a half times their rated capacity? ☐ ☐ ☐
    - Is a certificate of proof test available? ☐ ☐ ☐
    [29 CFR 1910.184(g)(5)]
14. Are metal mesh slings ONLY used below their permitted load rated capacities? ☐ ☐ ☐
    [29 CFR 1910.184(g)(6)]

Note: Consult the OSHA regulations or the manufacturer's requirements for permitted load ratings.

15. Are the sling manufacturer's recommendations followed concerning safe operating temperatures? ☐ ☐ ☐
    [29 CFR 1910.184(g)(7)]
16. Are all repairs to metal mesh slings performed by the manufacturer or another competent person? ☐ ☐ ☐
    [29 CFR 1910.184(g)(8)(i)]
17. Once repaired, are metal mesh slings marked or tagged, or are written records maintained to indicate the date and nature of the repair and the person or organization that performed the repairs? ☐ ☐ ☐
    [29 CFR 1910.184(g)(8)(ii)]
18. Are metal mesh slings immediately removed from service if any of the following conditions are present:
    - Broken weld or brazed joint is broken along the sling edge? ☐ ☐ ☐
    - Broken wire in any pat of the mesh? ☐ ☐ ☐
    - Reduction in wire diameter of 25 percent due to abrasion or 15 percent due to corrosion? ☐ ☐ ☐
    - Lack of flexibility due to distortion of the fabric? ☐ ☐ ☐
    - Distortion of the female handle so that the depth of the slot is increased by more than 10 percent? ☐ ☐ ☐
    - Distortion of either handles so that the width of the eye is decreased by more than 10 degrees? ☐ ☐ ☐
    - Reduction of 15 percent of the original cross-sectional areas of metal at any point around the handle eye? ☐ ☐ ☐
    - Distortion of either handle out of its plane? ☐ ☐ ☐
    - Cracked end fitting? ☐ ☐ ☐
    [29 CFR 1910.184(g)(9)]

6.2 Alloy Steel Chain Slings

	YES	NO	N/A

1. Do alloy steel chain slings have permanently affixed durable identification stating the size, grade, rated capacity, and reach?
[29 CFR 1910.184(e)(1) and 1926.251(b)(1)] ☐ ☐ ☐

2. Are manufacturer's recommended ratings used (when available) for safe working loads for the sizes of alloy steel chains and chain slings?
[29 CFR 1917.42(h)(1)] ☐ ☐ ☐

Note: When the manufacturer is unable to provide such ratings, MAKE SURE to use the tables for chains and chain slings found in American National Safety Standard for Slings, ANSI 30.9-1971.

3. Are riggers PROHIBITED to use
   - Proof coil steel chain (also known as common or hardware chain) and other chain not recommended by the manufacturer for slinging or hoisting? ☐ ☐ ☐
   - Kinked or knotted chains for lifting? ☐ ☐ ☐
   - Makeshift links or fasteners such as wire, bolts, or rods? ☐ ☐ ☐
   [29 CFR 1917.42(h)(2)]

4. Are riggers PROHIBITED to
   - Anneal alloy steel chains? ☐ ☐ ☐
   - Shorten chains by bolting, wiring, or knotting? ☐ ☐ ☐
   [29 CFR 1917.42(h)(6)]

5. Do hooks, rings, oblong links, pear-shaped links, welded or mechanical coupling links, or other attachments have rated capacities at least equal to that of the alloy steel chain with which they are used?
[29 CFR 1910.184(e)(2)(i) and 1926.251(b)(2)] ☐ ☐ ☐

6. Do ONLY designated, qualified persons inspect chains used for slinging and hoisting? ☐ ☐ ☐

7. Is a thorough inspection made of the alloy steel chain sling at least once every 12 months?
[29 CFR 1910.184(e)(3)(i) and 1926.251(b)(6)(i)] ☐ ☐ ☐

8. Are written records kept of these inspections?
[29 CFR 1910.184(e)(3)(ii) and 1926.251(b)(6)(ii)] ☐ ☐ ☐

9. Are thorough inspections of alloy steel chain slings, ☐ ☐ ☐
link by link for bent, stretched, cracks, or heat damage
in any link, performed by competent persons?
[29 CFR 1910.184(e)(3)(iii)]

Note: The inspector must check for wear, defective welds, deformation, and increase in length. OSHA construction regulations 1926.251(b)(5) require the sling be removed from service if any chain link has excessive wear. Consult OSHA regulation for additional details.

10. Are rings and hooks CHECKED daily for distortion, ☐ ☐ ☐
cracks in weld area, corrosion, and scores or heat
damage?
- Is the master link or are coupling links distorted? ☐ ☐ ☐
11. Do hooks have more than 15 percent greater than ☐ ☐ ☐
normal throat opening or more than 10 degrees twist
from plane of unbent hook?
- Are hook safety latches missing or nonfunctioning? ☐ ☐ ☐
12. Is the sling used at or below the rated capacity of the ☐ ☐ ☐
weakest component?
[29 CFR 1910.184(e)(2)(i)]
13. Are makeshift links or other fasteners formed from ☐ ☐ ☐
bolts or rods prohibited?
[29 CFR 1910.184(e)(2)(ii) and 1926.251(b)(3)]
14. Have new, repaired, or reconditioned alloy steel chain ☐ ☐ ☐
slings been proof-tested by the manufacturer, and is a
certificate of proof test available?
[29 [CFR 1910.184(e)(4) and (e)(7)(i)]
15. Are alloy steel chain slings used with loads at or below ☐ ☐ ☐
the rated capacities?
[29 CFR 1910.184(e)(5) and 1926.251(b)(4)]

Note: Consult the OSHA regulations or the manufacturer's requirements for permitted load ratings.

16. Are alloy steel chain slings permanently removed ☐ ☐ ☐
from service if heated above 1000°F?
[29 CFR 1910.184(e)(6)]
17. Are maximum working loads reduced in accordance ☐ ☐ ☐
with manufacturer's recommendations if the chain or
sling is exposed to temperatures above 600°F?
[29 CFR 1910(e)(6)]

18. Is the use of mechanical coupling links or low carbon steel repair links prohibited?
    [29 CFR 1910.184(e)(7)(ii)] ☐ ☐ ☐

19. Are chain slings removed from service when
    - Stretch has increased the length of a measured section by more than 5 percent? ☐ ☐ ☐
    - A link is bent, twisted, or otherwise damaged? ☐ ☐ ☐
    - A link has a raised scarf or defective weld? ☐ ☐ ☐
    - Maximum allowable wear is reached at any point of link (check OSHA tables)? ☐ ☐ ☐
    - Hooks are cracked or are opened more than 15 percent of the normal throat opening (measured at the narrowest point)? ☐ ☐ ☐
    - Hooks are twisted more than 10 degrees from the plane of the unbent hook?
      [29 CFR 1910.184(e)(9)(ii)] and
      [29 CFR 1910.184(e)(9)(ii)] ☐ ☐ ☐

20. Are chains repaired ONLY under qualified supervision? ☐ ☐ ☐

21. Are defective links or defective portions of chain links replaced with properly dimensioned links or connections of material similar to those of the original chain? ☐ ☐ ☐

22. Are repaired chains tested, before being returned to use, to the proof load recommended by the manufacturer for the original chain?
    [29 CFR 1917.42(h)(4)] ☐ ☐ ☐

Note: Tests MUST be performed by the manufacturer or certified by an agency accredited for the purpose. Test certificates MUST be available for inspection.

6.3 Natural Fiber Rope Slings and Hitches

Regulations

This checklist covers regulations issued by the U.S. Department of Labor, Occupational Safety and Health Administration (OSHA) under the general industry standard 29 CFR 1910.184 and the construction standard 1926.251. It applies to slings used with other equipment to move material by lifting or hoisting. The regulations cited apply only to private employers and their employees, unless adopted by a state agency and applied to other groups such as public employees.

Numerous tables in 29 CFR 1910.184 and 1926.251 give sling configuration, sling construction, sling diameter, and maximum load capacity. These tables have not been included as part of this checklist.
For additional information, consult the OSHA regulations.

	YES	NO	N/A
1. Are natural fiber rope slings inspected before each use?	☐	☐	☐

Note: Check with sling manufacturer for specific inspection recommendations.

	YES	NO	N/A
2. Are fiber rope slings ONLY made from new rope?	☐	☐	☐
▪ Is the use of repaired or reconditioned fiber rope slings prohibited? [29 CFR 1910.184(h)(6)]	☐	☐	☐
3. Are fiber rope slings that are made from conventional three-strand construction fiber rope used ONLY within their rated capacities and minimum diameter of curvature? [29 CFR 1910.184(h)(1)(i) and 1926.251(d)(1)]	☐	☐	☐

Note: Consult the tables in the OSHA regulations or the manufacturer's requirements for permitted load ratings and diameter of curvature restrictions for the different rigging situations. Diameter of curvature is important since wrapping a sling around something at a diameter less than that recommended reduces the strength of the sling at the bend and could cause failure of the line.

	YES	NO	N/A
4. Are natural fiber rope slings used ONLY within a temperature range of 20°F to 180°F? [29 CFR 1910.184(h)(2) and 1926.251(d)(3)]	☐	☐	☐
▪ Are the manufacturer's recommendations followed, for continued use, if natural fiber rope slings are used outside the temperature range of 20°F to 180°F or have been wetted or frozen? [29 CFR 1910.184(h)(2) and 1926.251(d)(3)]	☐	☐	☐
5. Is the use of spliced, fiber rope slings PROHIBITED unless they have been spliced in accordance with OSHA minimum requirements and with any additional recommendations of the manufacturer?	☐	☐	☐

Note: Consult the OSHA regulations under 29 CFR 1910.184(h)(3) and 1926.251(d)(2) and (4) for additional requirements on splices.

6. Do fiber rope slings have a minimum clear length of rope between eye splices equal to 10 times the rope diameter?
   [29 CFR 1910.184(h)(3)(iv) and 1926.251(d)(4)(iv)] ☐ ☐ ☐

7. Do eye splices consist of at least three full tucks? ☐ ☐ ☐

8. Do short splices consist of at least six full tucks, three on each side of the center line?
   [29 CFR 1917.42(c)(2)] ☐ ☐ ☐

9. Is the use of knots in place of splicing prohibited for fiber rope slings?
   [29 CFR 1910.184(h)(3)(v) and 1926.251(d)(2)(v)] ☐ ☐ ☐

10. Is the use of fiber rope slings prohibited if the end attachments in contact with the rope have sharp edges or projections?
    [29 CFR 1910.184(h)(4) and 1926.251(d)(5)] ☐ ☐ ☐

11. Are natural fiber rope slings immediately removed from service if any of the following conditions are present:
    - Abnormal wear? ☐ ☐ ☐
    - Powdered fiber between strands? ☐ ☐ ☐
    - Sufficient cut or broken fibers to affect the capability of the rope? ☐ ☐ ☐
    - Variations in the size or roundness of strands? ☐ ☐ ☐
    - Discolorations other than stains not associated with rope damage? ☐ ☐ ☐
    - Rotting? ☐ ☐ ☐
    - Distortion or other damage to hardware in the slings? ☐ ☐ ☐
    [29 CFR 1910.184(h)(5) and 1926.251(d)(6)]

12. Is each repaired sling proof-tested by the manufacturer (or another competent person or organization) to twice the rated capacity before its return to service? ☐ ☐ ☐
    - Is a certificate of proof test available? ☐ ☐ ☐
    [29 CFR 1910.184(i)(8)(ii)]

6.4 Synthetic Fiber Ropes and Web Slings

This checklist covers part of the regulations issued by OSHA under the General Industry standards 29 CFR 1910.184 and the Construction standards 1926.251. Numerous tables are included as part of 29 CFR 1910.184 which relates sling configuration, sling construction, sling diameter, and maximum load capacity. Consult the OSHA regulations.

Synthetic web slings must be inspected before use and should be removed from service if found to have acid or caustic burns, melting or charring of any part of the surface, snags, tears, or cuts, broken stitches, distorted fittings, or wear or elongation beyond the manufacturer's specifications.

SYNTHETIC FIBER ROPE SLINGS   YES   NO   N/A

1. Are synthetic fiber rope slings inspected before each use?   ☐   ☐   ☐

Note: Check with sling manufacturer for specific inspection recommendations.

2. Are synthetic fiber rope slings made from NEW rope, the ONLY type permitted?   ☐   ☐   ☐

3. Are riggers PROHIBITED to use temporarily repaired slings (including webbing and fittings)?
   [29 CFR 1910.184(i)(8)(iii)]   ☐   ☐   ☐

4. Are synthetic fiber rope slings used within a temperature range of 20°F to 180°F?
   [29 CFR 1910.184(h)(2)]   ☐   ☐   ☐

5. Has the manufacturer's recommendations been followed for continued use of synthetic fiber rope slings that have been wetted or frozen?
   [29 CFR 1910.184(h)(2)]   ☐   ☐   ☐

6. Are riggers PROHIBITED to use spliced synthetic fiber rope slings except when following manufacturer's recommendations?
   [29 CFR 1910.184(h)(2)]   ☐   ☐   ☐

7. Are riggers PROHIBITED to use knots in lieu of splicing for synthetic fiber rope slings?
   [29 CFR 1910.184(h)(3)(v)]   ☐   ☐   ☐

8. Are riggers PROHIBITED to use synthetic fiber slings if the end attachments in contact with the rope have sharp edges or projections?
   [29 CFR 1910.184(h)(4)]   ☐   ☐   ☐

9. Are synthetic fiber rope slings immediately removed from service if any of the following conditions are present:
   - Abnormal wear? ☐ ☐ ☐
   - Powdered fiber between strands? ☐ ☐ ☐
   - Sufficient cut or broken fibers to affect the capability of the rope? ☐ ☐ ☐
   - Variations in size or roundness of strands? ☐ ☐ ☐
   - Discolorations, other than stains, not associated with rope damage or rotting? ☐ ☐ ☐
   - Distortion or other damage to attached hardware? ☐ ☐ ☐
   [29 CFR 1910.184(i)(9)]

Synthetic Web Slings

10. Is each synthetic web slings marked or coated to show the rated capacity for each type of hitch and synthetic web material? ☐ ☐ ☐
    [29 CFR 1910.184(i)(1) and 1926.251(e)(1)]

11. Is the synthetic webbing of uniform thickness and width? ☐ ☐ ☐
    [29 CFR 1910.184(i)(2) and 1926.251(e)(3)]

Note: Selvage edges must not be split from the webbing's width.

12. Is each synthetic web sling marked or coated to show the rated capacity for each type of hitch and type of synthetic web material? ☐ ☐ ☐
    [29 CFR 1910.184(i)(1) and 1926.251(e)(1)]

13. Is the synthetic webbing of uniform thickness and width and selvage edges not split from the webbing's width? ☐ ☐ ☐
    [29 CFR 1910.184(i)(2)]

14. Are all fittings
    - Of a minimum breaking strength equal to that of the sling? ☐ ☐ ☐
    - Free of all sharp edges that could in any way damage the webbing? ☐ ☐ ☐
    [29 CFR 1910.184(i)(3)]

15. Is stitching the only method used to attach end fittings to webbing and to form eyes? ☐ ☐ ☐
    [29 CFR 1910.184(i)(4)]

16. Are synthetic web slings prohibited to be used with loads in excess of the rated capacity?
    [29 CFR 1910.184(i)(5) and 1926.251(e)(2)] ☐ ☐ ☐

17. When synthetic web slings are used, are the following precautions taken:
    - Nylon web slings PROHIBITED where fumes, vapors, sprays, mists, or liquids or acids or phenolics are present? ☐ ☐ ☐
    - Polyester and polypropylene web slings PROHIBITED where fumes, vapors, sprays, mists, or liquids of caustics are present? ☐ ☐ ☐
    - Web slings with aluminum fittings PROHIBITED where fumes, vapors, sprays, mists, or liquids of caustics are present? ☐ ☐ ☐
    [29 CFR 1910.184(i)(6)]

18. Are riggers PROHIBITED to use synthetic web slings of polyester and nylon slings at or above 180°F prohibited?
    [29 CFR 1910.184(i)(7)] ☐ ☐ ☐

19. Are riggers PROHIBITED to use polypropylene web slings at or above 200°F?
    [29 CFR 1910.184(i)(7)] ☐ ☐ ☐

20. Are synthetic web slings repaired ONLY by sling manufacturer or an equivalent entity?
    [29 CFR 1910.184(i)(8)(i)] ☐ ☐ ☐

21. Is each repaired sling proof-tested by the manufacturer or equivalent entity to twice the rated capacity prior to its return to service, and is a certificate of proof test available?
    [29 CFR 1910.184(i)(8)(ii)] ☐ ☐ ☐

22. Are riggers PROHIBITED to use temporarily repaired slings, including webbing and fittings?
    [29 CFR 1910.184(i)(8)(iii)] ☐ ☐ ☐

23. Are synthetic web slings immediately removed from service if any of the following conditions are present:
    - Acid or caustic burns? ☐ ☐ ☐
    - Melting or charring of any part of the sling surface? ☐ ☐ ☐
    - Snags, punctures, tears, or cuts? ☐ ☐ ☐
    - Broken or worn stitches? ☐ ☐ ☐

- Distortion of fittings? ☐ ☐ ☐
  [29 CFR 1910.184(i)(9)]

Note: Refer to other regulations, codes, and standards for additional information and safe operating practices, including: OSHA CFR 1910.184 and 1925.251 Regulations, Lift-All Catalog, ANSI/ASME B30.9, and the Web Sling and Tie Down Association Standards.

6.5 Standards (Consensus)

ANSI/ASME	American National Standards Institute/American Society of Mechanical Engineers
B30.9 2003	Slings
B30.10 2008	Hooks
B30.20 2008	Below-the-Hook Devices
CI	Cordage Institute
	Rope test methods, standards, and guidelines
WRTB	(Wire Rope Technical Board)
	Wire Rope Users Manual
	Wire Rope Slings User's Manual
WSTDA	(Web Sling Tie Down Association)
RS-1	Synthetic Polyester Round Slings
TH-1	Synthetic Sewing Threads for Slings and Tie Downs
WS-1	Synthetic Web Slings
WB-1	Synthetic Webbing for Slings

Chapter 7

Rigging Hardware

7.1 Overhead Lifting Systems

Rigging is only as strong as its weakest link. It doesn't matter what safe working load (SWL) is stamped on a hook. If the hook is cracked and twisted or opening up at the throat, it cannot deliver its full-rated capacity.

Only load-rated hardware of forged alloy steel should be used for overhead lifting. Load-rated hardware is stamped with its working load limit (WLL).

Adequate capacity is the first thing to look for in rigging hardware. For overhead lifting, the design factors MUST be 5:1. Once the right hardware has been chosen for a job, it has to be inspected regularly as long as it's in service. Hardware that has been weakened in use should be replaced.

A person trained in compliance with the schedule in ANSI B30.10 should conduct a visual periodic inspection for cracks, nicks, wear, gouges, and deformation as part of a comprehensive documented inspection program that covers rigging hardware in use or stored on site.

Maintenance guidelines

Use a good checklist when performing preventive maintenance to make sure all the work items are performed, and make notes of any issues that should be checked into further.

- Schedule repairs or replacements if any of the following are identified:
  - Cracks—Inspect closely; some cracks are very fine.
  - Missing parts—Make sure that parts suck as catches on hooks, nuts on cable clips, and cotter pins in shackle pins are still in place.
  - Stretching—Check hooks, shackles, and chain links for signs of opening up, elongation, and distortion.

- Periodically inspect hooks used in frequent load cycles or pulsating loads.

Note: Hook and threads should be periodically inspected by magnetic particle or dye penetrant. Some disassembly may be required.

- A designated, qualified person MUST inspect all new, altered, modified, or repaired rigging hardware before use—to verify compliance with ANSI/ASME B30.10. (A written record, however, is NOT required.)

Note: Hardware inspection depends on environmental conditions, application, storage of product prior to use, frequency of use, etc.; when in doubt, inspect hardware prior to each use.

- Rigging equipment operator or a designated, qualified person MUST perform a visual inspection each day before hardware is used.
- Carefully check each item for wear, deformation, cracks, or elongation—sure sign of imminent failure.
- Withdraw immediately from service any damaged hardware or hardware having any condition that may result in a hazard. Repair or replace such units; however, written records are NOT required for frequent inspections.
- Designated, qualified person MUST periodically perform a complete inspection of hardware, examining for cracks, missing parts, stretching, and distortion, and determining whether conditions constitute a hazard.

Note: Periodic inspections MUST NOT exceed 1 year.

- Base frequency of inspections on
  - Frequency of hardware use
  - Severity of service conditions
  - Nature of lifts being made
  - Experience gained on service life of the hardware
- Determine time intervals on
  - Normal service—yearly
  - Severe service—monthly to quarterly
  - Special service—as recommended by a qualified person
- KEEP a written record of the most recent periodic inspection and include condition of the hardware to help pinpoint problems and to ensure periodic inspection intervals.

- Destroy, rather than discard, any hardware that has been judged to be defective to prevent its being used again by someone not aware of the hazard involved.
- Remove rigging hardware from service if any of the following damages is visible:
  - Missing or illegible manufacturer's name or trademark and rated load identification of both
  - Indications of heat damage, including weld splatter or arc strikes
  - Excessive pitting or corrosion
  - Bent, twisted, distorted, stretched, elongated, cracked, or broken load-bearing components.
  - Reduction of 10 percent of the original or catalog dimension at any point around the body or pins
  - Incomplete thread engagement
  - Excessive thread damage
  - Evidence of unauthorized modifications
  - Other conditions, including visible damage, that cause doubt as to the continuous use of the equipment
  - Lack of ability for swivel hoist rings to freely rotate or pivot
  - Loose or missing, for swivels, nuts, bolts, cotter pins, snap rings, or other fasteners and retaining devices
- DO NOT return rigging hardware to service unless approved by a qualified person.

Source: Brookhaven National Laboratory, Brookhaven, New York.

REGULATIONS	YES	NO	N/A

1. Are all overhead lifting systems inspected each day before being used for damage or defects by a competent person designated by the employer? ☐ ☐ ☐

2. Are additional inspections performed during sling use, where service conditions are warranted?
[29 CFR 1926.251(a)(6)] ☐ ☐ ☐

Note: Damaged or defective fastenings and attachments MUST be immediately removed from service.

3. Are overhead lifting systems NOT loaded in excess of its recommended safe working load? (See Tables H-1 through H-20, 1926.252(e) for the specific equipment.)
[29 CFR 1926.251(a)(2)] ☐ ☐ ☐

4. Are overhead lifting systems, when not in use, removed from the immediate work area so as not to present a hazard to employees?
   [29 CFR 1926.251(a)(3)]

5. Are special custom design grabs, hooks, clamps, or other lifting accessories, marked to indicate the safe working loads for such units as modular panels, prefabricated structures, and similar materials?
   - Are they proof-tested prior to their use to 125 percent of their rated load?
     [29 CFR 1926.251(a)(4)]

7.2 Attachments, Fittings, and Connections

Although there are a wide variety of connections, fittings, and end attachments for rigging applications, only forged alloy steel, load-rated types should be used for overhead lifting and hoisting to ensure the highest degree of safety. All rigging hardware should have their SWL stamped directly on them.

Maintenance guidelines

Use a good checklist when performing preventive maintenance to make sure all the work items are performed, and make notes of any issues that should be checked into further.

Connections	Fittings	Attachments
Eyes	Collets	Hooks
Ferrules	Sockets	Rings and links
Mechanical splices	Thimbles	Eye and ring bolts
Swivels	Clips (clamps)	Turnbuckles/bottle screws
	Shackles (clevises)	Spreader/equalizing beams
		Overhauling weights

- MAKE SURE a competent person, designated by the employer, each day before use inspects slings and all connections, fittings, and attachments—in accordance with manufacturer specifications.
- Check slings for rust and broken wires—cut up or mark defective ropes as UNUSABLE
- Perform additional inspections during sling use where service conditions are warranted.
- Remove damaged or defective slings and hardware immediately from service.

Rigging Hardware 149

- MAKE SURE rope clips attached with U-bolts have the U-bolts on the dead end or short end of the rope.
- Verify that hooks, rings, oblong links, pear-shaped links, welded or mechanical coupling links, or other attachments, when used with alloy steel chains, have a rated capacity at least equal to that of the chain.
- Ensure all eye splices are made in an approved manner with rope thimbles. (Sling eyes are excepted.)
- Check that positive latching devices are used to secure loads.
- Ensure all custom-designed lifting accessories are marked to indicate their safe working loads.
  - Lifting accessories are proof-tested to 125 percent of their rated load.
- Are all rigging systems properly stored?

Wire rope clips (clamps)
- Check wire rope clips to ensure they are permanently and legibly marked with the size and manufacturer's identification mark.
- Verify that clips meet or exceed the requirements of ASME B30.26 (R 2009)—Rigging Hardware.
- MAKE SURE clamps are legibly and permanently marked with size and the manufacturer's identification mark.
- Assemble clamps using the same size, type, class, and manufacturer, and follow manufacturer's instructions for proper installation.
- Visually inspect clamps before use for damage, corrosion, wear, and cracks.
- Verify that clamps are legibly and permanently marked in accordance with OSHA 29 CFR 1926.251(c)(5).
- Ensure that the assembled clamp contains the same size, type, and class parts.

Eyebolts
- Use only shouldered eyebolts for rigging hardware, except when prohibited by the configuration of the item to which the eyebolt is attached.
- Use non-shouldered eyebolts, when required, only in vertical pulls or in rigging systems that are designed, analyzed, and approved by a qualified person.

Note: Eyebolts designed for and permanently installed by the manufacturer on existing engineered equipment are considered part of the engineered equipment and are acceptable for their intended use—only as long as they pass visual inspection before use.

- MAKE SURE shoulders seat uniformly and snugly against the surface on which they bear.

Note: DO NOT use 7/8-in eyebolts because a 7/8- to 9-in Unified National Coarse (UNC) thread may be threaded into a 1- to 8-in UNC tapped hole that will fail when loaded.

- Use eyebolts for hoisting that are only made of
  - Forged carbon steel—with the manufacturer's name or identification mark forged in raised characters on the surface of the eyebolt, or
  - Alloy steel—with the symbol "A" (denoting alloy steel) and manufacturer's name or identification mark forged in raised characters on the surface of the eyebolt

Note: Alloy steel eyebolts are forged, quenched, and tempered with improved toughness properties, intended primarily for low-temperature applications.

- Check each eyebolt carefully (mandatory) before using.
  - Visually check the hole to ensure that there has been NO deformation.
  - Check the condition of the threads in the hole to ensure that the eyebolt will secure and the shoulder can be brought down snug.
  - Ensure that the shank of the eyebolt is not undercut and is smoothly radiused into the plane of the shoulder or the contour of the ring for non-shouldered eyebolts.
  - Ensure that eyebolts are provided with a jam nut of a type that does not depend upon deformation of the threads for security.
- Ensure a minimum thread shank length of engagement to attain the rated capacity for threaded hole applications:
  - Steel—1 thread diameter
  - Cast iron, brass, bronze—one and a half times the thread diameter
  - Aluminum, magnesium, zinc, plastic—two times the thread diameter
- Destroy any eyebolts that are cracked, bent, or have damaged threads.
- MAKE SURE that all eyebolts have a minimum design factor of 5 based on ultimate strength—in accordance with ANSI/ASME B18.15—Forged Eyebolts.

Note: Regardless if eyebolts from selected manufacturers have a higher SWL, they MUST HAVE a design factor of NO less than 5—verified before use.

- DO NOT use nuts, washers, and drilled plates or assemble them to make shouldered eyebolts.
- DO NOT use wire type or welded eyebolts (or both) in lifting operations.
- Inspect and evaluate all manufacturer-installed lift points before use for cracks, deformation, excessive wear, or damage.

Note: Consult equipment custodian or cognizant engineer when questions arise regarding the use of manufacturer-installed lift points.

Swivel hoist rings

- Check that these attachments are marked to clearly identify UNC or metric threads, the manufacturer, the safe working load, and the torque value.
- Verify availability of manufacturer's instructions for swivel hoist rings.
- Ensure free movement of the swivel bail—360 degrees swivel and 180 degrees pivot.
- MAKE SURE the work surface is flat and smooth to provide flush seating for the bushing flange.
- Tighten the bolt to full torque loading.

Note: Loosening of a swivel hoist ring bolt may develop after prolonged service in a permanent installation. Periodically verify proper torque; retighten the mounting bolt as recommended by the manufacturer. In lieu of other direction from the manufacturer, check hoist swivel ring mounting bolts for proper torque before each lift.

- MAKE SURE if swivel hoist ring is installed with a retention nut, that the nut has no less than full thread engagement and that it is torqued in accordance with the manufacturer's recommendations.
- NEVER use free-fit spacers or washers between the swivel host ring bushing flange and the mounting surface.
- NEVER use swivel hoist rings that show signs of corrosion, wear, or damage.
- Read, understand, and follow the manufacturer's instructions, diagrams, and chart information before using a swivel hoist ring.

- VERIFY eyebolts and swivel hoist rings are
  - Uniform in quality consistent with good manufacturing and inspection practices
  - Free from imperfections which, resulting from their nature, degree, or extent, would make the eyebolt or swivel hoist ring unsuitable for the intended use
- NEVER exceed the SWL of a swivel hoist ring as specified by the manufacturer.

Source: Pacific Northwest National Laboratory PNNL, Richland, WA.

END ATTACHMENTS	YES	NO	N/A
1. Are welding of end attachments, except covers to thimbles, performed prior to the assembly of the sling? [29 CFR 1926.251(c)(15)(i)]	☐	☐	☐
2. Are all welded end attachments NOT used unless proof-tested by the manufacturer or equivalent entity at twice their rated capacity prior to initial use? [29 CFR 1926.251(c)(15)(ii)]	☐	☐	☐

Note: aa certificate of proof test MUST be maintained and made available for examination.

3. Are fiber rope slings NOT used if end attachments, in contact with the rope, have sharp edges or projections? [29 CFR 1926.251(d)(5)]	☐	☐	☐

Fittings

4. Are all rigging system fittings			
• Of a minimum breaking strength equal to that of the sling? [29 CFR 1926.251(e)(4)(i)]	☐	☐	☐
• Free of all sharp edges that could in any way damage the webbing? [29 CFR 1926.251(e)(4)(ii)]	☐	☐	☐
5. Is stitching the ONLY method used to attach end fittings to webbing and to form eyes?	☐	☐	☐
• Is the thread in an even pattern, and does it contain a sufficient number of stitches to develop the full breaking strength of the sling? [29 CFR 1926.251(e)(5)]	☐	☐	☐

6. Are eyebolts loaded ONLY in the plain of the eye? ☐ ☐ ☐
   - Are they NOT loaded at angles of less than 45 degrees to the horizontal? ☐ ☐ ☐
7. Are shoulderless eyebolts NOT loaded at an angle? ☐ ☐ ☐
8. Is compression hardware (wire rope clamps) CHECKED for
   - Sufficient number of wire rope clips? ☐ ☐ ☐
   - Properly tightened wire rope clips? ☐ ☐ ☐
   - Indications of damaged wire rope? ☐ ☐ ☐
   - Indications of slippage of wire rope? ☐ ☐ ☐

7.3 Hooks and Shackles

HOOKS   YES   NO   N/A

1. Are the manufacturer's recommendations followed in determining the safe working loads of the various sizes and types of specific and identifiable hooks? ☐ ☐ ☐
   - Are ONLY properly marked shackles used? ☐ ☐ ☐
2. Are all hooks, for which no applicable manufacturer's recommendations are available, tested to twice the intended safe working load before they are initially put into use? ☐ ☐ ☐
3. Are records maintained of the dates and results of such tests? ☐ ☐ ☐
   [29 CFR[1926.251(f)(2)]

Note: Table H-19 [29 CFR 1926.251(f)(1)] shall be used to determine the safe working loads of various sizes of shackles, except that higher safe working loads are permissible when recommended by the manufacturer for specific, identifiable products, provided that a safety factor of not less than 5 is maintained.

4. Are hooks or shackles checked for any sign of distortion? ☐ ☐ ☐

Note: A bent pin or a distorted hook or shackle is no longer safe.

5. Are hooks NEVER used whose throat opening has been increased, or shows a tip has been bent more than 10 degrees out of plane from hook body, or in any other way distorted or bent? ☐ ☐ ☐

Note: A latch will not work properly on a hook with a bent or worn tip.

6. Are bent or sprung hooks discarded? ☐ ☐ ☐
   [29 CFR 1917.42(j)(2)]

7. Are loads applied ONLY to the throat of the hook? ☐ ☐ ☐
   [29 CFR 1917.42(j)(5)]

8. Are workers PROHIBITED from using a hook that ☐ ☐ ☐
   is worn beyond its safe limits?

9. Is any hook removed from service if it has a crack, ☐ ☐ ☐
   nick, or gouge?

Note: Hooks with a crack, nick, or gouge MUST be repaired by grinding lengthwise, following the contour of the hook, provided that the reduced dimension is within the prescribed limits.

10. Are workers PROHIBITED from repairing, altering, ☐ ☐ ☐
    reworking, or reshaping a hook by welding, heating,
    burning, or bending?

11. Are workers PROHIBITED from side loading, back ☐ ☐ ☐
    loading, or tip loading a hook?

12. Are eye hooks, shank hooks, and swivel hooks used ☐ ☐ ☐
    ONLY with wire rope or chain?

Note: Efficiency or assembly may be reduced when used with synthetic material.

13. Are workers PROHIBITED from swiveling the ☐ ☐ ☐
    swivel hook while it is supporting a load?

14. Is the use of a latch mandated by regulations or safety ☐ ☐ ☐
    codes (such as OSHA, MSHA, ANSI/ASME B30,
    insurance carriers, and others)?

Note: When using latches, check manufacturer's WARNINGS.

15. Does the hook ALWAYS support the load (the latch ☐ ☐ ☐
    MUST NEVER support the load)?

16. Is the angle from the vertical to the outermost leg, ☐ ☐ ☐
    when placing two sling legs in hook, NEVER
    greater than 45 degrees?
    - Does the angle between the legs NOT exceed ☐ ☐ ☐
      90 degrees?

Note: For angles greater than 90 degrees or more than two legs, a master link or bolt-type anchor shackle MUST be used to attach the legs of the sling to the hook.

SHACKLES

	YES	NO	N/A
1. Are the manufacturer's recommended safe working loads for shackles NOT exceeded? [24 CFR 1917.42(i)(1)]	☐	☐	☐

Note: In the absence of manufacturer's recommendations, OSHA 1917.42 Table C-3 shall apply.

	YES	NO	N/A
Screw pin shackles used aloft in house fall or other gear, except in cargo hook assemblies, shall have their pins moused or otherwise effectively secured. [29 CFR 1917.42(i)(2)]	☐	☐	☐

7.4 Sheaves, Blocks, and Tackles

SHEAVES

	YES	NO	N/A
1. Are sheaves compatible with the size of rope used, as specified by the manufacturer?	☐	☐	☐
2. Are sheaves inspected to ensure they are of correct size, properly aligned, lubricated, and in good condition?	☐	☐	☐
3. Are sheaves equipped with cable-keepers when rope is subject to riding or jumping off a sheave?	☐	☐	☐

BLOCKS

	YES	NO	N/A
1. Are blocks disassembled periodically and inspected and lubricated?	☐	☐	☐
▪ Are bearings and bushings properly lubricated to help prevent wearing?	☐	☐	☐
▪ Are shells properly preserved to reduce deterioration?	☐	☐	☐
2. Is the block's frame checked for any cracks or splits and for any signs of the sheave wearing on the inside of the frame?	☐	☐	☐
▪ Is the pin checked to see if it is bent?	☐	☐	☐

Note: Dropping a wooden block can split its frame.

	YES	NO	N/A
3. Are workers PROHIBITED from painting a wooden block; a coat of paint could hide a split?	☐	☐	☐
4. Is ONLY clear shellac or varnish or several coats of linseed oil used to coat wood blocks?	☐	☐	☐

Note: Inspect and replace any suspected wooden blocks.

5. Is the metal on wooden blocks frequently and carefully inspected for any signs of distortion or wear? ☐ ☐ ☐

Note: Metal in constant use is subject to fatigue. Immediately replace any doubtful block.

6. Are wire metal rope blocks, in continuous use, disassembled frequently and inspected for wear? ☐ ☐ ☐

Note: Metal rope blocks, used only occasionally, seldom need to be disassembled if they are kept well lubricated. Check before replacing an entire block, to see if a replacement part is available for any part that is defective.

7.5 Below-the-Hook Lifting Devices

Below-the-hook lifting devices include spreader bars, lifting yokes, lifting baskets, and lift fixtures.

- Verify that rated capacity of each lifting device is marked on the main structure where it is visible and legible.

Note: If the lifting device comprises several items, each detachable from the assembly, each lifting device MUST be marked with its rated capacity.

- MAKE SURE, at a minimum, that a nameplate, name tag, or other permanent marker is affixed to below-the-hook lifting devices displaying the following data:
  - Manufacturer's name
  - Serial number or unit identification
  - Weight of lifting-device—if over 100 lb (45.36 kg)
  - Electric power (when applicable)
  - Pressure and volume of compressed air (when applicable)
  - Rated capacity
  - Proof of inspection label by hoist and rigging inspector
- Relabel lifting devices that are re-rated with the new rated capacity.

Note: If a lifting device cannot be marked with its rated capacity and weight, it MUST, however, be marked with an identification number, and its documentation MUST describe its rated capacity and its weight.

- Affix a label (or labels) to each vacuum lifting device in a readable position to display the word WARNING or other legend designed to bring the label to the attention of the operator.
- Include information on the label that cautions against
  - Exceeding the rated capacity or lifting loads not specified in the manufacturer's instruction manual
  - Operating a damaged or malfunctioning unit or a unit with missing parts
  - Incorrect positioning of the lifting device on the load
  - Lifting people
  - Moving loads above people
  - Removing/obscuring warning labels
  - Operating the lifting device when the rated capacity, lifting-device weight, or safety markings are missing
  - Making alterations or modifications to the lifting device.
  - Lifting loads higher than necessary and leaving suspended loads unattended.
- Affix a label to each unit that directs the user to consult manufacturer's manual if the size or shape of the unit prohibits inclusion of the preceding markings.

7.6 Standards (Consensus)

ANSI/ASME

B30.9 (2006)	Slings
B30.10 (1999)	Hooks
B30.20 (1985)	Below-the-Hook Lifting Devices
B30.26 (R2009)	Rigging Hardware—Safety Standard for rigging hardware used for lifting purposes, such as shackles, links, rings, swivels, turnbuckles, eyebolts, hoist rings, wire rope clips, wedge sockets, and rigging blocks
B30.27 (2009)	Material Placement Systems—Safety Standard for cableways, cranes, derricks, hoists, hooks, jacks, and slings—applies to material placement systems, commonly referred to as concrete pumping machines

Chapter 8

Scaffolding Systems

Besides problems with planks and guardrails, the main causes of injuries and deaths on scaffolds are poor planning for assembling and taking them apart, failing to use them properly, missing tie-ins or bracing, loads that are too heavy, and being too close to power lines—as well as falling objects that can hurt people below scaffolds.

Additionally, equipment failure at attachment points, and parts failure resulting from overloading of scaffolding, as well as changing environmental conditions (high winds, temperature extremes or the presence of toxic gases).

To ensure safe scaffold erection and use begin by developing policy and work rules. Sources of information for policy development and work rules include OSHA and ANSI standards, scaffold trade associations, scaffolding suppliers, and safety and engineering consultation services.

8.1 Guidelines for Setup, Use, and Maintenance

Scaffold setup

All scaffolding systems MUST be installed and maintained by employees who have been specifically trained in this task.

Temporary devices such as boxes or chairs are PROHIBITED on the scaffold, to allow workers to reach higher heights.

	YES	NO	N/A
1. Are scaffolds erected, assembled, altered, moved, or taken apart under the direction of someone competent in scaffold erection?	☐	☐	☐

Inspections

- Are alteration and dismantling activities planned and performed with the same care as with erection? ☐ ☐ ☐
- Are any incomplete scaffolds or damaged components tagged out of service? ☐ ☐ ☐

2. Are the footings of a standing scaffold on a
   - Sound and rigid foundation? ☐ ☐ ☐
   - NOT set on soft ground, frozen ground that could melt? ☐ ☐ ☐
   - NOT resting on blocks? ☐ ☐ ☐
   - With base plates attached to feet, extending at least 1 ft (30.48 cm) past each leg? ☐ ☐ ☐
   - Capable of carrying four times the maximum intended load? ☐ ☐ ☐

3. Is the scaffold level? ☐ ☐ ☐

4. Are poles, legs or uprights of scaffolds plumb and securely braced to prevent swaying and displacement? ☐ ☐ ☐

5. Is a scaffold that is more than four times higher than its base widely tied to supports? ☐ ☐ ☐

6. Is the platform complete front to back and side to side, fully planked or decked with
   - NO gaps greater than 1 in (25.4 mm) between planks and uprights? ☐ ☐ ☐
   - Platforms and walkways 18 in (45.72 cm) wide or more? ☐ ☐ ☐
   - Planking of sufficient stress grade or scaffold grade timber? ☐ ☐ ☐
   - Wood planks UNPAINTED so that any cracks will show? ☐ ☐ ☐
   - Planks 10 ft (3.05 m) long extended at least 6 in (15.24 cm) past the end supports, but NOT more than 12 in (30.48 cm), or secured from movement? ☐ ☐ ☐

7. Are guardrails and toeboards provided on scaffolds more than 10 ft (3 m) above ground? ☐ ☐ ☐

8. Are all sections pinned or appropriately secured? ☐ ☐ ☐

9. Is there a safe way to get on and off the scaffold, such as a ladder, without climbing on cross braces? ☐ ☐ ☐

10. Is the front face, for scaffolds, within 14 in (35.56 cm) of the work—for outrigger scaffolds within 3 ft (91.44 cm)? ☐ ☐ ☐

11. Does the scaffold meet following electrical safety clearance distances:
    - 10 ft (3.05 m) or more from power lines? ☐ ☐ ☐
    - 3 ft (91.44 cm) if lines are less than 300 V— unless lines are first de-energized? ☐ ☐ ☐
12. Are workers PROHIBITED from using unstable objects, such as barrels, boxes, bricks, or blocks, to support scaffolds or planks? ☐ ☐ ☐
13. Are scaffolds secured to permanent structures with anchor bolts or other means? ☐ ☐ ☐
14. Is planking of sufficient stress grade or scaffold grade timber? ☐ ☐ ☐
15. Does planking extend over end supports between 6 in (15.24 cm) and 18 in (45.72 cm)? ☐ ☐ ☐
16. Are multiple planks and platforms either overlapped NOT LESS than 6 in (15.24 cm) or NOT MORE than 12 in (30.48 cm)—or are they secured to ensure stability? ☐ ☐ ☐
17. Are guardrails, midrails, or toe boards installed on the open sides and ends of platforms that are more than 4 ft (1.219 cm) above the ground or floor level? ☐ ☐ ☐
18. Are guardrails provided on scaffolds more than 10 ft (3.086 m) above the ground? ☐ ☐ ☐
19. Are scaffolds equipped with toe boards wherever there is a possibility that falling material could cause a hazard? ☐ ☐ ☐
20. Are toe boards at least 4 in (10.16 cm) in height? ☐ ☐ ☐

Scaffold use

No employees may erect or work on scaffolding systems that are covered with ice, snow, or any other slippery substance.

Working during high winds is only permissible if employees wear personal fall protection systems and protect themselves with suitable wind screens.

When using scaffolds, MAKE SURE suspension ropes are kept from contact with sources of heat (welding, cutting, etc.), acids, and other corrosive substances.

All employees working more than 10 ft (3 m) above the ground must be protected from falls using a system of guardrails or a protective harness.

A system of canopies, toe boards, and guardrails may be used to protect people walking below. Protective systems normally are designed during the permit process, with help from local officials.

	YES	NO	N/A
1. Does a competent person, trained in scaffold safety, inspect each scaffold before each work shift and after anything happens that could affect the structure?	☐	☐	☐
2. Is the competent person trained in scaffold safety?	☐	☐	☐
3. Are defective parts on scaffolds immediately replaced or repaired?	☐	☐	☐
4. Are guardrails or personal fall arrest systems (or both) used if a work area is less than 18 in (45.71 cm) wide?	☐	☐	☐
5. Are guardrails and toe boards installed on all open sides and ends of scaffold platforms?	☐	☐	☐
6. Are railings protecting floor openings, platforms, or scaffolds equipped with toe boards wherever there is a possibility that falling material could cause a hazard?	☐	☐	☐
7. Are toe boards at least 4 in (10.2 cm) in height?	☐	☐	☐
8. Are hardhats worn by all workers on and around the scaffold?	☐	☐	☐
9. Is personal fall protection provided if the scaffold is over 10 ft (3.05 cm) high, or are guardrails over 38 in (91.44 cm) high installed?	☐	☐	☐
10. Are lifelines used for scaffolds that are more than 10 ft (3.086 m) off the ground?	☐	☐	☐
11. Are lifelines firmly anchored to an overhead structure and not to the scaffold when workers are working on suspended scaffolds?	☐	☐	☐
12. Are workers wearing body harnesses attached to lifelines?	☐	☐	☐
13. Are workers PROHIBITED from loading scaffolds in excess of their maximum load limits?	☐	☐	☐
14. Are heavy loads placed over or near the bearers and not in the center of the plank?	☐	☐	☐

15. Are damaged scaffolds or defective parts on a scaffold immediately replaced or repaired? ☐ ☐ ☐
16. Where persons work under scaffold, is a 0.5-in (12.7-mm) mesh screen provided between toe board and guardrail? ☐ ☐ ☐
17. Is ice or snow removed from scaffolds, and is sand applied to the wood, before conducting work in winter weather? ☐ ☐ ☐
18. Are heavy tools, equipment, and supplies hoisted up, rather than carried up by hand? ☐ ☐ ☐
19. Are scaffold loads (including tools and other equipment)
    - Kept to a minimum? ☐ ☐ ☐
    - NOT allowed to accumulate on scaffolds? ☐ ☐ ☐
    - Removed when the scaffold is not in use (such as at the end of a day)? ☐ ☐ ☐
20. Is debris and unnecessary material removed from scaffold platforms that could cause slipping and falls? ☐ ☐ ☐
21. Are materials secured before moving a scaffold? ☐ ☐ ☐
22. Are employees removed from the scaffold before it is moved? ☐ ☐ ☐
23. Are scaffolds dismantled and removed when they are no longer needed? ☐ ☐ ☐

Note: DO NOT use temporary scaffolding as a permanent installation.

24. Is there a way to get on or off a scaffold—if it is more than 2 ft (60.96 cm) above or below a level—such as a ladder, ramp, or personnel hoist? ☐ ☐ ☐
25. Is the access NOT greater than 14 in (35.66 cm) from the scaffold? ☐ ☐ ☐
26. Are scaffold loads (including tools and other equipment) kept to a minimum and removed when the scaffold is not in use (like at the end of a day)? ☐ ☐ ☐
27. Are employees removed from scaffolds and PROHIBITED from working on scaffolds during high winds, rain, snow, or bad weather? ☐ ☐ ☐
28. Are scaffold suspension ropes kept from contact with sources of heat (welding, cutting, etc.) and from acids and other corrosive substances? ☐ ☐ ☐

Sources: CPWR—Center to Protect Workers' Rights, AFL-CIO Building and Construction Trades Department, Oklahoma State University, Stillwater, OK.

Maintenance guidelines

Use a good checklist when performing preventive maintenance to make sure all the work items are performed, and make notes of any issues that should be checked into further.

Follow all scaffolding manufacturer's recommendations—as well as all federal, state, and local regulations, codes, and ordinances pertaining to scaffolds and scaffolding systems.

- Inspect scaffold to ensure it has not been altered and is in safe working condition.
- DO NOT dismantle or alter scaffold systems unless under the supervision of a qualified person.
- Continuously inspect erected scaffolds to ensure that they are maintained in a safe condition.
  - Check for loose fittings, such as bolts, nuts, or clamps.
  - Report any unsafe condition to a supervisor.
- Prevent damage to scaffolding systems by exposure to corrosive or other damaging substances.
- Clean and service scaffolding systems regularly to remove buildup of dirt, paint, or grime.
- DO NOT use harsh chemicals that may cause metal to rust.
  - If a pressure washer is used, thoroughly wipe scaffolding down to reduce the chances of rust or rotting.
  - If there is rust or paint already present, use a power sander or sandpaper to remove it.

Note: Cleaning not only will protect and preserve the life of scaffolding equipment, but also will enable following local safety and code precautions.

- Take inventory of any bent, broken, loose, or missing pieces when breaking down scaffolding.
- Label each piece with duct tape, painter's tape, or some other color-coded system (electrical tape comes in a variety of colors)—making it easy to keep all similar pieces together, like screws of a certain length.
- Record all maintenance work done on scaffolds.
- Implement a system to track information concerning necessary repairs and parts that need to be replaced.

- Make a special space for tools, so they will be easily accessible during assembly and breaking down.
- Provide proper storage of scaffolding equipment.
- Group items by type (all the planks, all the bolts) or keep the pieces of each scaffolding unit together.
- Install utility shelving for large pieces, and use bins, cubbies, tool racks, and drawers for smaller pieces; a place for bolts and screws is just as important as a place for poles and planks.

Note: A storage area does not have to be large or complicated. Cleaning, maintaining, repairing, organizing, and storing scaffolding equipment should make maintenance work easier and safer, not more difficult.

8.2 Suspended Scaffolds

All suspended scaffolding systems must be designed by an engineer who is experienced in this type of structural design.

The system should be installed by a competent person and inspected before each use. Only items, designed as counterweights, may be used for this purpose, and must be secured mechanically to the scaffolding system.

All support items such as hooks, roof irons, and parapet clamps must be made of steel or iron and supported by tiebacks that have the same strength as the scaffold's ropes.

No gas-powered hoists of lift systems may be used on suspended scaffolds.

1. Provide proper training for all workers who use any type of suspended scaffold or fall protection equipment.
2. Follow scaffold manufacturers' guidance regarding the assembly, rigging, and use of scaffolds.
3. Comply with the current and proposed OSHA regulations for working with scaffolds.
4. Ensure that design and construction of scaffolds conform with OSHA requirements.
5. Supporting outrigger beams MUST be able to support at least four times the intended load.

Note: To keep a scaffold from falling to the ground, it must be attached to the roof, tied to a secure anchorage, or secured with counterweights. The suspension ropes and rigging MUST support at least six times the intended load.

6. Attach counterweights to secure and strong places on a building so they won't move.

Note: DO NOT use bags of sand or gravel, masonry blocks, or roofing materials that can flow or move.

7. DO NOT use gas-powered equipment or hoists.
8. Hoists MUST have automatic brakes for emergencies.
9. Suspended scaffolds (1 or 2 points) MUST be tied or secured to prevent swaying.
10. Hoists MUST have automatic brakes for emergencies.
11. Shield scaffold suspension ropes and body belt or harness system droplines (lifelines) from hot or corrosive processes, and protect them from sharp edges or abrasion.
12. Inspect all scaffolds, scaffold components, and personal fall protection equipment before each use.
13. Provide personal fall protection equipment and make sure that it is used by all workers on suspended scaffolds.
14. Use structurally sound portions of buildings or other structures to anchor droplines for body belt or harness systems and tiebacks for suspended scaffold support devices.
15. Secure droplines and tiebacks to separate anchor points on structural members.

Source: NIOSH.

8.3 Stationary Scaffolds

All non-suspended scaffolds, with a height-to-base width ratio greater than 4 to 1, MUST be braced or supported to prevent tipping.

Ties and braces MUST be installed at both ends of a scaffolding system, and also at intervals of every 30 ft (9.1 m) horizontally.

A system of bracing MUST be used on any size ratio system when the scaffold will be subjected to unequal side loads, such as cantilevered platforms.

All vertical members on supported scaffolds must be set onto stable base plates and must not be placed on loose objects or on mechanical equipment.

When scaffolds are assembled, inspected, or dismantled, a qualified person—having the experience and training to identify existing and predictable hazards—MUST be in charge of the operation on-site to supervise. The qualified person is responsible for inspecting each scaffold prior to its use at the beginning of each shift or new workday.

Scaffolds MUST be erected and dismantled according to design standards, engineered specifications, or manufacturer's instructions, and wood scaffolds must be designed by a licensed engineer.

REGULATIONS	YES	NO	N/A
1. Are scaffolds erected ONLY on solid footing or have secure anchorage? [29 CFR 1910.28(a)(1)]	☐	☐	☐
2. Are workers PROHIBITED from using unstable objects, such as barrels, boxes, loose bricks, or concrete blocks to support scaffolds or planks? [29 CFR 1910.28(a)(2)]	☐	☐	☐
3. Are scaffolds, and their components, sound, rigid, and sufficient to carry their own weight plus four times the maximum intended load without settling or displacement? [29 CFR 1910.28(a)(4)]	☐	☐	☐
4. Are scaffolds maintained in a safe condition?	☐	☐	☐
5. Are workers PROHIBITED from altering or moving scaffolds horizontally while they are in use or occupied? [29 CFR 1910.28(a)(5)]	☐	☐	☐
6. Are any scaffolds damaged or weakened from any cause immediately repaired and NOT used until repairs have been completed? [29 CFR 1910.28(a)(6)]	☐	☐	☐
7. Are scaffolds NEVER loaded in excess of the working load for which they are intended? [29 CFR 1910.28(a)(7)]	☐	☐	☐
8. Are all load-carrying timber members of scaffold framing a minimum of 1500 f (stress grade) construction grade lumber? [29 CFR 1910.28(a)(8)]	☐	☐	☐

Note: All dimensions are nominal sizes as provided in the American Lumber Standards (ALS), except that where rough sizes are noted, only rough and undressed lumber or the size specified will satisfy minimum requirements. Where nominal sizes of lumber are used in place of rough sizes, the nominal-size lumber shall be such as to provide equivalent strength to that specified in the ALS Tables D-7 through D-12 and D-16.

168   Inspections

9. Is all planking Scaffold Grade as recognized by grading rules for the species of wood used? (Check OSHA tables for maximum permissible spans for 2- × 9-in or wider planks.)
[29 CFR 1910.28(a)(9)] ☐ ☐ ☐

10. Do scaffold planks extend over their end supports neither less than 6 in (15.24 cm) nor more than 18 in (45.72 cm)?
[29 CFR 1910.28(a)(13)] ☐ ☐ ☐

11. Is all planking or are platforms overlapped a minimum 12 in (30.48 cm) or secured from movement?
[29 CFR 1910.28(a)(11)] ☐ ☐ ☐

12. Are nails or bolts used in the construction of scaffolds of adequate size and in sufficient numbers at each connection to develop the designed strength of the scaffold? ☐ ☐ ☐

Note: Nails MUST NOT be subjected to a straight pull and MUST be driven full length.

13. Are the poles, legs, or uprights of scaffolds plumb and securely and rigidly braced to prevent swaying and displacement?
[29 CFR 1910.28(a)(14)] ☐ ☐ ☐

14. Are scaffolds erected, moved, dismantled, or altered under the supervision of a competent person? ☐ ☐ ☐

15. Is an access ladder or equivalent safe access provided?
[29 CFR 1910.28(a)(12)] ☐ ☐ ☐

16. Is overhead protection provided for workers on a scaffold exposed to overhead hazards?
[29 CFR 1910.28(a)(16)] ☐ ☐ ☐

17. Are scaffolds equipped with guardrails, midrails, and toe boards? ☐ ☐ ☐

18. Are scaffolds provided with a screen between the toe board and the guardrail, extending along the entire opening, consisting of No. 18 gauge U.S. Standard Wire 0.5-in (12.7-mm) mesh or the equivalent, where persons are required to work or pass under the scaffolds?
[29 CFR 1910.28(a)(17)] ☐ ☐ ☐

19. Are scaffold accessories such as braces, brackets, trusses, screw legs or ladders that are damaged or weakened by any cause immediately repaired or replaced? ☐ ☐ ☐
20. Does a "competent person" inspect scaffolding and, at designated intervals, reinspect it? ☐ ☐ ☐
21. Is rigging on suspension scaffolds inspected by a competent person before each shift and after any occurrence that could affect structural integrity to ensure that all connections are tight and that no damage to the rigging has occurred since its last use? ☐ ☐ ☐
22. Is synthetic and natural rope used in suspension scaffolding protected from heat-producing sources? ☐ ☐ ☐
23. Are employees instructed about the hazards of using diagonal braces as fall protection? ☐ ☐ ☐
24. Do workers access scaffolds only by using ladders and stairwells? ☐ ☐ ☐
25. Are scaffolds erected at least 10 ft (3.05 m) from electric power lines at all times? ☐ ☐ ☐
26. Do materials being hoisted onto a scaffold have a tag line?
[29 CFR 1910.28(a)(15)] ☐ ☐ ☐
27. Are workers PROHIBITED from working on scaffolds during storms or high winds?
[29 1910.28(a)(18)] ☐ ☐ ☐
28. Are workers PROHIBITED from working on scaffolds that are covered with ice or snow, unless all ice or snow is removed and planking sanded to prevent slipping?
[29 1910.28(a)(19)] ☐ ☐ ☐
29. Are workers PROHIBITED from allowing tools, materials, and debris to accumulate in quantities to cause a hazard?
[29 1910.28(a)(20)] ☐ ☐ ☐
30. Is only treated or protected fiber rope used for or near any work, involving the use of corrosive substances or chemicals?
[29 CFR 1910.28(a)(21)] ☐ ☐ ☐
31. Is wire or fiber rope used for scaffold suspension capable of supporting at least six times the intended load?
[29 CFR 1910.28(a)(22)] ☐ ☐ ☐

32. Are ONLY wire rope scaffolds used when acid ☐ ☐ ☐
solutions are used for cleaning buildings over 50 ft
(15.24 m) in height?
[29 CFR 1910.28(a)(23)]

33. Are scaffolds secured to permanent structures, ☐ ☐ ☐
through use of anchor bolts, reveal bolts, or other
equivalent means?
[29 CFR 1910.28(a)(26)]

Note: Window cleaners' anchor bolts MUST NOT be used.

34. Are special precautions taken to protect scaffold ☐ ☐ ☐
members, including any wire or fiber ropes, when
using a heat-producing process?
[29 CFR 1910.28(a)(27)]

8.4 Mobile Scaffolds

Regulations

29 CFR 1910.29   Safety Requirements for Manually Propelled Mobile Ladder Stands and Scaffolds (Towers)

	YES	NO	N/A
1. Are all exposed surfaces of mobile ladder stands free from sharp edges, burrs, or other safety hazards? [29 CFR 1910.29(a)(2)(v)]	☐	☐	☐
2. Is the maximum work level height less than or equal to four times the minimum or least base dimension of the mobile ladder stand?	☐	☐	☐
• Is the base dimension at least one-half the height—if workers are riding the scaffolding? [29 CFR 1910.29(a)(3)(i)]	☐	☐	☐

Note: Suitable outrigger frames may be used to achieve the required base dimension or other means used to guy or brace the unit against tipping.

3. Is the minimum step width for ladder stands 16 in ☐ ☐ ☐
(40.64 cm)?
[29 CFR 1910.29(a)(3)(ii)]

4. Are the steps of ladder stands fabricated from ☐ ☐ ☐
slip-resistant treads?
[29 CFR 1910.29(a)(3)(iv)]

5. Are at least two of the four casters equipped with a swivel lock to prevent movement?
[29 CFR 1910.29(a)(4)(ii)] □ □ □

6. Are steps of mobile ladder stands uniformly spaced?
[29 CFR 1910.29(f)(3)] □ □ □

7. Are steps of mobile ladder stands sloped, with a rise that is not less than 9 in (22.86 cm) and not more than 10 in (25.4 cm) and a depth of at least 7 in (17.78 cm)?
[29 CFR 1910.29(f)(3)] □ □ □

Note: The slope of the steps section shall be a minimum of 55 degrees and a maximum of 60 degrees from the horizontal.

8. Are mobile ladder stands with more than five steps equipped with handrails?
[29 CFR 1910.29(f)(4)(i)] □ □ □

9. Are the handrails at least 29 in (73.66 cm) high?
[29 CFR 1910.29(f)(4)(ii)] □ □ □

Note: Measurements must be taken vertically from the center of the steps.

10. Are all ladder stands with a work level 10 ft (3.05 cm) or higher above the ground or floor equipped with a standard, 4-in (10.16-cm) nominal toe board?
[29 CFR 1910.29(a)(3)(vi)] □ □ □

Sources: NIOSH—National Institute for Occupational Safety and Health, CDC Atlanta, GA.

8.5 Aerial Lifts (Platforms)

INDUSTRY GUIDELINES   **YES NO N/A**

1. Are workers who operate aerial lifts properly trained in the safe use of the equipment? □ □ □

2. Are elevating work platforms maintained and operated in accordance with manufacturer's instructions? □ □ □

3. Are workers PROHIBITED to overriding hydraulic, mechanical, or elevated electrical safety devices? □ □ □

4. Are workers PROHIBITED from moving the equipment with workers in any elevated platform—unless permitted by the manufacturer? ☐ ☐ ☐

5. Are workers PROHIBITED from positioning themselves between overhead hazards, such as joists and beams, and the rails of the basket? ☐ ☐ ☐

Note: Movement of the lift could crush the workers(s).

6. Is a minimum clearance, of at least 10 ft (3 m) away from the nearest overhead lines, ALWAYS maintained? ☐ ☐ ☐

7. Do workers ALWAYS treat power lines, wires, and other conductors as energized—even if they are down or appear to be insulated? ☐ ☐ ☐

8. Do workers ALWAYS use a body harness or restraining belt with a lanyard attached to the boom or basket to prevent being ejected or pulled from the basket? ☐ ☐ ☐

9. Are brakes ALWAYS set—and are wheel chocks used—when equipment is on an incline? ☐ ☐ ☐

9. Are outriggers used, if provided? ☐ ☐ ☐

10. Are workers PROHIBITED from exceeding load limits of the equipment? ☐ ☐ ☐

Note: Allow for the combined weight of the worker, tools, and materials.

8.6 Construction Scaffolds

Construction scaffolding includes a variety of temporary structures erected at worksites for construction, alteration, demolition, or repair work, including painting and decorating.

REGULATIONS

This checklist covers regulations issued by the U.S. Department of Labor, Occupational Safety and Health Administration (OSHA), under 29 CFR 1926.451, Subpart L, of the construction standards.

The checklist does not address criteria for suspension scaffolds, suspension ropes, and stair towers. Consult OSHA Standard 29 CFR 1926.451 for these types of scaffolds.

The regulations cited apply only to private employers and their employees, unless adopted by a state agency and applied to other groups such as public employees.

This checklist should be used with the scaffolding checklist—Fall Protection Systems.

Subpart L of the OSHA construction standards includes appendixes that provide useful information on scaffold specifications.

	YES	NO	N/A
1. Does each scaffold and scaffold component support (without failure) its own weight and at least four times the maximum intended load? [29 CFR 1926.451(a)(1)]	☐	☐	☐

Note: The stall load of any scaffold hoist must not exceed three times its rated load [29 CFR 1926.451(a)(5)]. Appendix A of the OSHA regulations gives directions for constructing scaffolds.

	YES	NO	N/A
2. Are all working levels on scaffolds fully planked or decked between the front uprights and the guardrail supports? [29 CFR 1926.451(b)(1)]	☐	☐	☐
3. Are scaffold platform spaces 1 in (25.4 mm) or less between adjacent units and uprights? [29 CFR 1926.451(b)(1)(i)]	☐	☐	☐

Note: Spaces up to 9.5 in (24.13 cm) wide are permitted around uprights. If platforms are used only as walkways or during erecting or dismantling, the employer may establish the space between planking as necessary to provide safe working conditions.

	YES	NO	N/A
4. Are all scaffold platforms at least 18 in (45.72 cm) wide? [29 CFR 1926.451(b)(2)]	☐	☐	☐

Note: Scaffold platforms less than 18 in (45.72 cm) wide are permitted if wider platforms are not possible. Workers who use these platforms, however, MUST be protected by guardrails or personal fall arrest systems.

	YES	NO	N/A
5. Is the distance between the front edge of the scaffold platform and the face of the work 14 in (35.56 cm) or less, unless guardrail systems are put along the front edge, or personal fall arrest systems are used? [29 CFR 1926.451(b)(3)]	☐	☐	☐

Note: Exceptions are permitted under special situations. Consult OSHA regulations for details. Maximum distance from the face for plastering and lathing operations is 18 in (45.72 cm).

6. Does the end of each scaffold platform extend over the centerline of its support at least 6 in (15.24 cm) unless cleated or otherwise restrained by hooks or equivalent means?
   [29 CFR 1926.451(b)(4)]   ☐ ☐ ☐

7. If the scaffold platform is 10 ft (3.05 m) or less in length, does the end of the scaffold platform extend 12 in (30.48 cm) or less over its support?
   [29 CFR 1926.451(b)(5)(i)]   ☐ ☐ ☐

Note: The end of the scaffold platform may extend more than 12 in (30.48 cm) over its support if the platform is designed and installed so that the cantilevered portion of the platform can support workers or materials without tipping, or a guardrail blocks worker access to the cantilevered end.

8. On scaffolds where planks are abutted to form a long platform, does each plank end rest on a separate support surface?
   [29 CFR 1926.451(b)(6)]   ☐ ☐ ☐

Note: Common support members, such as "T" sections, may be used to support abutting planks. Hook-on platforms designed to rest on common supports may also be used.

9. On scaffolds where platforms overlap to form a long platform, does the overlap occur over supports?   ☐ ☐ ☐
   - Is the overlap at least 12 in (30.48 cm) unless the platform is nailed together or otherwise restrained to prevent movement?
     [29 CFR 1926.451(b)(7)]   ☐ ☐ ☐

10. At points of a scaffold where the platform changes direction, is this procedure followed:
    - Lay the platform that rests on a bearer at an angle other than a right angle?   ☐ ☐ ☐
    - Lay the platform that rests at right angles over the same bearer, on top of first platform?   ☐ ☐ ☐
      [29 CFR 1926.451(b)(8)]

11. Are workers PROHIBITED from covering wood platforms on scaffolds with opaque finishes?
    [29 CFR 1926.451(b)(9)]   ☐ ☐ ☐

Note: Platform edges may be covered or marked for identification. Wood platforms may be coated periodically with wood preservatives, fire-retardant finishes, and slip-resistant finishes; however, the coating MUST NOT obscure the top or bottom wood surfaces.

12. Do scaffold components from different manufacturers fit together without force and maintain the scaffold's structural integrity?
[29 CFR 1926.451(b)(10)] ☐ ☐ ☐

13. Are scaffold components from different manufacturers modified ONLY if a competent person determines that the scaffold made of the "mixed" parts is structurally sound?
[29 CFR 1926.451(b)(10)] ☐ ☐ ☐

14. Are scaffold components made of dissimilar metals used together ONLY if a competent person has determined that galvanic action will not reduce the strength of any component to an unacceptable level?
[29 CFR 1926.451(b)(11)] ☐ ☐ ☐

SUSPENDED SCAFFOLDS YES NO N/A

1. Are supporting outrigger beams able to support at least four times the intended load? ☐ ☐ ☐

2. Is the scaffold securely attached to the roof, tied to a secure anchorage, or secured with counterweights? ☐ ☐ ☐

3. Do the suspension ropes and rigging support at least six times the intended load? ☐ ☐ ☐

4. Are counterweights attached to secure, strong places on a building, so they won't move? ☐ ☐ ☐

Note: DO NOT use bags of sand or gravel, masonry blocks, or roofing materials that can flow or move.

5. Is the use of gas-powered equipment or hoists PROHIBITED for scaffolds? ☐ ☐ ☐

6. Are 1- or 2-point suspended scaffolds tied or secured to prevent swaying? ☐ ☐ ☐

SUPPORTED SCAFFOLDS YES NO N/A

1. Are supported scaffolds with a height-to-base width ratio of more than 4 to 1 restrained from tipping by guying, tying, bracing, or equivalents?
[29 CFR 1926.451(c)(1)] ☐ ☐ ☐

Note: Install guys, ties, and braces at locations where horizontal members support both inner and outer legs. Install guys, ties, and braces according to the scaffold manufacturer's recommendations or at the closest horizontal member to the 4 to 1 height. Repeat vertically at locations of horizontal members every 20 ft (6.09 m) or less thereafter for scaffolds 3 ft (91.44 cm) wide or less, and every 26 ft (7.93 m) or less thereafter for scaffolds greater than 3 ft (91.44 cm) wide. Place the top guy, tie, or brace of completed scaffolds no further than the 4 to 1 height from the top. Install guys, ties, and braces at each end of the scaffold and at horizontal intervals 30 ft (9.144m) or less [measured from one end (not both) toward the other]. Use ties, guys, braces, or outriggers to prevent tipping when there is an eccentric load, such as a cantilevered work platform.

2. Are supported scaffold poles, legs, posts, frames, and uprights placed on base plates and mud sills or other firm foundation?
   [29 CFR 1926.451(c)(2)]

3. Are footings level, sound, and rigid?
   - Can they support the loaded scaffold without settling or displacement?
   [29 CFR 1926.451(c)(2)(i)]

4. Are workers PROHIBITED from using unstable objects for supporting scaffolds and platform units?
   [29 CFR 1926.451(c)(2)(ii)]

5. Are workers PROHIBITED from using unstable objects as working platforms?
   [29 CFR 1926.451(c)(2)(iii)]

6. Are workers PROHIBITED from using front-end loaders and similar pieces of equipment to support scaffold platforms, unless they were designed by the manufacturer for such use?
   [29 CFR 1926.451(c)(2)(iv)]

7. Are supported scaffold poles, legs, posts, frames, and uprights plum and braced to prevent swaying and displacement?
   [29 CFR 1926.451(c)(3)]

Wood Pole Scaffolds

8. When platforms are moved to the next level, are existing platforms left undisturbed until the new bearers have been set in place and braced?
   [29 CFR 1926.452(a)(1)]

9. Is crossbracing installed between the inner and outer set of poles on double-pole scaffolds?
[29 CFR 1926.452(a)(2)]
☐ ☐ ☐

10. Is diagonal bracing that is installed in both directions across the entire inside face of double-pole scaffolds used to support loads equivalent to a uniformly distributed load of 50 psf (244.13 kg/cm^2) or more per square foot?
[29 CFR 1926.452(a)(3)]
☐ ☐ ☐

11. Is diagonal bracing installed in both directions across the entire outside face of all double- and single-pole scaffolds?
[29 CFR 1926.452(a)(4)]
☐ ☐ ☐

12. Are runners and bearers installed on edge?
[29 CFR 1926.452(a)(5)]
☐ ☐ ☐

13. Do bearers extend at least 3 in (7.62 cm) over the outside edges of the runners?
[29 CFR 1926.452(a)(6)]
☐ ☐ ☐

14. Do runners extend at least over two poles? ☐ ☐ ☐
    - Are they supported by bearing blocks that are securely attached to the poles? ☐ ☐ ☐
[29 CFR 1926.452(a)(7)]

15. Are workers PROHIBITED from splicing braces, bearers, and runners between poles?
[29 CFR 1926.452(a)(8)]
☐ ☐ ☐

16. If wooden poles are spliced, are they square? ☐ ☐ ☐
    - Does the upper section rest squarely on the lower section? ☐ ☐ ☐
[29 CFR 1926.452(a)(9)]

Note: Wood splice plates must be on at least two adjacent sides, and must extend at least 2 ft (60.96 cm) on either side of the splice, overlap the abutted ends equally, and have at least the same cross-sectional areas as the pole. Splice plates of other materials of equivalent strength may be used.

Tube-and-Coupler Scaffolds

17. When platforms are moved to the next level, are existing platforms left undisturbed until the new bearers have been set in place and braced?
[29 CFR 1926.452(b)(1)]
☐ ☐ ☐

18. Are transverse braces that form an "X" across the width of the scaffold installed at the scaffold ends and at least at every third set of posts horizontally (measured form one end) and every fourth runner vertically?
[29 CFR 1926.452(b)(2)] ☐ ☐ ☐

19. Does bracing extend diagonally from the inner or outer posts or runners upward to the next outer or inner posts or runners?
[29 CFR 1926.452(b)(2)] ☐ ☐ ☐

20. Are building ties installed at the bearer levels between he transverse bracing? [29 CFR 1926.452(b)(2)] ☐ ☐ ☐

21. On straight run scaffolds, is longitudinal bracing across the inner and outer rows of posts installed diagonally in both directions? ☐ ☐ ☐
    - Does bracing extend from the base of the end posts upward to the top of the scaffold at approximately a 45-degree angle?
    [29 CFR 1926.452(b)(3)] ☐ ☐ ☐

Note: On scaffolds whose length is greater than their height, such bracing must be repeated beginning at least at every fifth post. On scaffolds whose length is less than their height, such bracing must be installed from the base of the end posts upward to the opposite end posts, and then in alternating directions until reaching the top of the scaffold. Bracing must be installed as close as possible to the intersection of the bearer and post or runner and post.

22. If bracing cannot be attached to posts, is it attached to the runners as close to the post as possible?
[29 CFR 1926.452(b)(4)] ☐ ☐ ☐

23. Are bearers installed transversely between posts? ☐ ☐ ☐
    - When coupled to the posts, does inboard coupler bear directly on the runner coupler?
    [29 CFR 1926.452(b)(5)]

Note: When the bearers are coupled to the runners, the couplers must be as close to the posts as possible.

24. Do bearers extend beyond the posts and runners, and provide full contact with the coupler?
[29 CFR 1926.452(b)(6)] ☐ ☐ ☐

25. Are runners installed along the length of the scaffold, on both the inside and outside posts at level heights?
[29 CFR 1926.452(b)(7)] ☐ ☐ ☐

Note: When tube and coupler guardrails and midrails are used on outside posts, they may be used in place of outside runners.

26. Are runners interlocked on straight runs to form continuous lengths, and coupled to each post?
[29 CFR 1926.452(b)(8)] ☐ ☐ ☐

Note: Install bottom runners and bearers as close to the base as possible.

27. Are couplers made of structural metal, such as drop-forged steel, malleable iron, or structural grade aluminum?
[29 CFR 1926.452(b)(9)] ☐ ☐ ☐

Note: The use of gray case iron is prohibited.

Fabricated Frame Scaffolds

28. When moving platforms to the next level, are existing platforms left undisturbed until the new end frames have been set in place and braced?
[29 CFR 1926.452(c)(1)] ☐ ☐ ☐

29. Are frames and panels braced by cross, horizontal, or diagonal braces (or combinations thereof), to secure vertical members together laterally?
[29 CFR 1926.452(c)(2)] ☐ ☐ ☐

Note: The crossbraces shall be long enough to automatically square and align vertical members so that the erected scaffold is plumb, level, and square. All brace connections must be secured.

30. Are frames and panels joined together vertically by coupling or stacking pins or equivalent means?
[29 CFR 1926.452(c)(3)] ☐ ☐ ☐

31. Are frames and panels locked together vertically by pins or equivalent means at points where uplift could displace scaffold end frames or panels?
[29 CFR 1926.452(c)(4)] ☐ ☐ ☐

Bricklayers' Square Scaffolds

32. Are wood scaffolds reinforced with gussets on both sides of each corner?
[29 CFR 1926.452(e)(1)] ☐ ☐ ☐

33. Are diagonal braces installed on all sides of each square?
[29 CFR 1926.452(e)(2)] ☐ ☐ ☐

34. Are diagonal braces installed between squares on the rear and front sides of the scaffold, and do they extend from the bottom of each square to the top of the next square?
[29 CFR 1926.452(e)(3)]   ☐ ☐ ☐

35. Are scaffolds three tiers or less in height? Are they constructed so that one square rests directly above the other?
[29 CFR 1926.452(e)(4)]   ☐ ☐ ☐

Note: The upper tiers shall stand on a continuous row of planks laid across the next lower tier, and shall be nailed down or otherwise secured to prevent displacement.

Horse Scaffolds

36. Are scaffolds less than 10 ft (3.05 m) high?   ☐ ☐ ☐
    - Are they two tiers high or less?   ☐ ☐ ☐
      [29 CFR 1926.452(f)(1)]

37. When horses are arranged in tiers, is each horse placed directly over the horse in tier below?
[29 CFR 1926.452(f)(2)]   ☐ ☐ ☐

38. When horses are arranged in tiers, are the legs of each horse nailed down or otherwise secured to prevent displacement?
[29 CFR 1926.452(f)(3)]   ☐ ☐ ☐

39. When horses are arranged in tiers, is each tier crossbraced?
[29 CFR 1926.452(f)(4)]   ☐ ☐ ☐

Ladder Jack Scaffolds

40. Are all ladder jack scaffolds 20 ft (6.1 m) or less above the ground?
[29 CFR 1926.452(k)(1)]   ☐ ☐ ☐

41. Are ladders that are used to support ladder jack scaffolds in compliance with OSHA regulations?
(See subpart X of this part—Stairways and Ladders.)
[29 CFR 1926.452(k)(2)]   ☐ ☐ ☐

42. Are ladder jacks designed and constructed so that they will bear on the side rails in addition to the ladder rungs?
[29 CFR 1926.452(k)(3)]   ☐ ☐ ☐

Note: If bearing is on rungs only, the bearing area shall be at least 10 in on each rung.

43. Are ladders that are used to support ladder jacks placed, fastened, or equipped with devices to prevent slipping?
[29 CFR 1926.452(k)(4)] ☐ ☐ ☐

44. Are workers PROHIBITED from bridging scaffold platforms one to another?
[29 CFR 1926.452(k)(5)] ☐ ☐ ☐

TRAINING YES NO N/A

1. Are workers trained to recognize the hazards associated with the type of scaffold used and to understand the procedures to control or minimize those hazards?
[29 CFR 1926.454(a)] ☐ ☐ ☐

Note: Consult the OSHA regulations for the specific training areas that must be covered.

2. Are trainers qualified in the subject matter?
[29 CFR 1926.454(a)] ☐ ☐ ☐

3. Are workers who erect, disassemble, move, operate, repair, maintain, or inspect a scaffold trained to recognize hazards associated with the work? ☐ ☐ ☐
   - Are the trainers who train the workers competent?
   [29 CFR 1926.454(b)] ☐ ☐ ☐

Note: Consult the OSHA regulations for the specific training areas that must be covered.

4. Are workers retrained when they show a lack of skills or understanding needed for safe work, involving erecting, using, or dismantling scaffolds?
[29 CFR 1926.454(c)] ☐ ☐ ☐

8.7 Scaffold Access

REGULATIONS YES NO N/A

1. Are ladders, stairs, ramps, or walkways provided to access scaffold platforms more than 2 ft (60.96 cm) above or below a point of access?
[29 CFR 1926.451(e)(1)] ☐ ☐ ☐

Note: Crossbraces must not be used as a means of access.

2. Are portable, hook-on, and attachable ladders positioned to prevent scaffold from tipping?
[29 CFR 1926.451(e)(2)(i)] ☐ ☐ ☐

3. Are hook-on and attachable ladders positioned so the bottom rung is not more than 24 in (60.96 cm) above the scaffold supporting level?
[29 CFR 1926.451(e)(2)(ii)] ☐ ☐ ☐

4. Are hook-on and attachable ladders designed for the scaffold in use?
[29 CFR 1926.451(e)(2)(iv)] ☐ ☐ ☐

5. Do hook-on and attachable ladders have rung length of at least 11.5 in (29.21 cm)?
[29 CFR 1926.451(e)(2)(v)] ☐ ☐ ☐

6. Do hook-on and attachable ladders have uniformly spaced rungs with a maximum spacing between rungs of 16.75 in (42.55 cm)?
[29 CFR 1926.451(e)(2)(vi)] ☐ ☐ ☐

7. Is the bottom step of stairway-type ladders 24 in (60.96 cm) or less above the scaffold supporting level?
[29 CFR 1926.451(e)(3)(i)] ☐ ☐ ☐

8. Do stairway-type ladders have rest platforms at 12 ft (3.66 m) maximum vertical intervals?
[29 CFR 1926.451(e)(3)(ii)] ☐ ☐ ☐

9. Do stairway-type ladders have a step width of at least 16 in (40.64 cm)?
[29 CFR 1926.451(e)(3)(iii)] ☐ ☐ ☐

Note: Mobile scaffold stairway-type ladders may have a minimum step width of 11.5 in (29.21 cm).

10. Do stairway-type ladders have slip-resistant treads on all steps and landings?
[29 CFR 1926.451(e)(3)(iv)] ☐ ☐ ☐

11. Do ramps and walkways 6 ft (1.83 m) or more above lower levels have guardrails?
[29 CFR 1926.451(e)(5)(i)] ☐ ☐ ☐

12. Are ramps and walkways inclined with a slope less than 1 vertical to 3 horizontal 20 degrees above the horizontal?
[29 CFR 1926.451(e)(5)(ii)] ☐ ☐ ☐

13. Do ramps and walkways that are steeper than □ □ □
    1 vertical in 8 horizontal have cleats 14 in (35.56 cm),
    or less, apart that are securely fastened to the planks
    to provide footing?
    [29 CFR 1926.451(e)(5)(iii)]

14. Are integral prefabricated scaffold access frames □ □ □
    constructed for use as ladder rungs?
    [29 CFR 1926.451(e)(6)(i)]

15. Do integral prefabricated scaffold access frames have □ □ □
    rung lengths of at least 8 in (20.32 cm)?
    [29 CFR 1926.451(e)(6)(ii)]

16. Do workers have fall protection if integral □ □ □
    prefabricated scaffold access frames with rungs less
    than 11.5 in (29.21 cm) are used as work platforms?
    [29 CFR 1926.451(e)(6)(iii)]

17. Are integral prefabricated scaffold access frames □ □ □
    uniformly spaced within each frame section?
    [29 CFR 1926.451(e)(6)(iv)]

18. Do integral prefabricated scaffold access frames have □ □ □
    a maximum spacing between rungs of 16.75 in
    (42.55 cm)?
    [29 CFR 1926.451(e)(6)(v)]

Note: Nonuniform rung spacing, caused by joining end frames together, is allowed, provided the resulting spacing is 16.75 in (42.55 cm) or less.

19. Do steps and rungs of ladder and stairway-type □ □ □
    access line up vertically with each other between
    rest platforms?
    [29 CFR 1926.451(e)(7)]

20. Is the horizontal distance 14 in (35.56 cm), or less, □ □ □
    and the vertical distance 24 in (60.96 cm), or less,
    between two surfaces used to provide direct access
    between them?
    [29 CFR 1926.451(e)(8)]

21. During erecting and dismantling of supported □ □ □
    scaffolds, does a competent person provide and
    evaluate safe means of access?
    [29 CFR 1926.451(e)(9)(i)]

22. During erecting and dismantling of supported scaffolds, are hook-on or attachable ladders installed as soon as they can be used safely?
[29 CFR 1926.451(e)(9)(ii)] ☐ ☐ ☐

23. During erecting and dismantling of supported scaffolds, are the ends of tubular welded frame scaffolds used as climbing devices for access only if the horizontal members are parallel, level, and 22 in (55.99 cm) apart, or less, vertically?
[29 CFR 1926.451(e)(9)(iii)] ☐ ☐ ☐

24. During erecting and dismantling of supported scaffolds, is it prohibited to use the crossbraces on tubular welded frame scaffolds for access or exit?
[29 CFR 1926.451(e)(9)(iv)] ☐ ☐ ☐

SCAFFOLD USE   YES   NO   N/A

1. Are scaffolds and scaffold components loaded below their maximum intended loads or rated capacities (whichever is less)?
[29 CFR 1926.451(f)(1)] ☐ ☐ ☐

2. Is the use of shore or lean-to scaffolds prohibited?
[29 CFR 1926.451(f)(2)] ☐ ☐ ☐

3. Does a competent person inspect scaffolds and scaffold components for visible defects before each work shift, and after any occurrence that could affect a scaffold's structural integrity?
[29 CFR 1926.451(f)(3)] ☐ ☐ ☐

4. Are parts of a scaffold that are damaged or weakened immediately repaired, replaced, braced, or removed from service until repaired?
[29 CFR 1926.451(f)(4)] ☐ ☐ ☐

5. Is the horizontal movement of a scaffold PROHIBITED while workers are on the scaffold (unless the scaffold is designed for movement by a registered professional engineer, or is a mobile scaffold meeting OSHA standards)?
[29 CFR 1926.451(f)(5)] ☐ ☐ ☐

6. Are proper clearances (as shown in Tables 8.1 and 8.2) between scaffolds and power lines always maintained?
[29 CFR 1926.451(f)(6)] ☐ ☐ ☐

TABLE 8.1 Insulated Power Lines

Insulated line voltage	Minimum distance	Alternative
Less than 300 V	3 ft (91.44 cm)	
300 V to 50 kV	10 ft (3.1 m)	
More than 50 kV	10 ft (3.1 m) plus 4.0 in (10 cm) for each 1 kV over 50 kV	Two times the length of the line insulator, but never less than 10 ft (3.1 m)

Note: Scaffolds and materials may be closer to power lines if such clearance is necessary, and only after the utility company or electrical system operator has been notified, and the utility company or electrical system operator has de-energized the lines, relocated the lines, or installed protective coverings to prevent contact with the lines.

7. Are scaffolds erected, moved, dismantled, or altered only under the supervision and direction of a competent person qualified in scaffold erection, moving, dismantling, or alteration? □ □ □
[29 CFR 1926.451(f)(7)]

8. Are scaffolds erected, moved, dismantled, or altered only by experienced and trained employees selected for such work by the competent person? □ □ □
[29 CFR 1926.451(f)(7)]

9. Are workers PROHIBITED from working on scaffolds covered with snow, ice, or other slippery material, except as necessary to remove such materials? □ □ □
[29 CFR 1926.451(f)(8)]

10. If swinging loads are hoisted onto or near scaffolds, are tag lines or equivalent measures used to control the loads? □ □ □
[29 CFR 1926.451(f)(9)]

TABLE 8.2 Uninsulated Power Lines

Uninsulated line voltage	Minimum distance	Alternative
Less than 50 kV	10 ft (3.1 m)	
More than 50 kV	10 ft (3.1 m) plus 4.0 in (10 cm) for each 1 kV over 50 kV	Two times the length of the line insulator, but never less than 10 ft (3.1 m)

11. Is working on scaffolds during storms or high winds prohibited unless a competent person has determined that it is safe for workers to be on the scaffold and workers are protected by a personal fall arrest system or wind screens?
[29 CFR 1926.451(f)(12)] ☐ ☐ ☐

12. Is debris removed from platforms?
[29 CFR 1926.451(f)(13)] ☐ ☐ ☐

13. Are makeshift devices, such as boxes and barrels, prohibited on scaffold platforms for increasing the working level height?
[29 CFR 1926.451(f)(14)] ☐ ☐ ☐

14. Is it prohibited to use ladders on scaffolds to increase the working level height?
[29 CFR 1926.451(f)(15)] ☐ ☐ ☐

Note: Ladders may be used on large-area scaffolds if certain conditions are met. Consult OSHA regulations for the required conditions.

15. Are scaffold platforms used only if they deflect 1/60 of the span (or less) when loaded?
[29 CFR 1926.451(f)(16)] ☐ ☐ ☐

8.8 Guardrail Systems, Canopies, and Covers

REGULATIONS YES NO N/A

1. Are other structural members (such as additional midrails and architectural panels) installed so that openings in the guardrail system are 19 in (48.26 cm) wide or less?
[29 CFR 1926.502(b)(2)(iv)] ☐ ☐ ☐

2. Can guardrail systems withstand (without failure) a force of at least 200 lb (90.718 kg) applied within 2 in (5.08 cm) of the top edge, in any outward or downward direction, at any point along the top edge?
[29 CFR 1926.502(b)(3)] ☐ ☐ ☐

3. When a 200 lb (90.718 kg) test load is applied in a downward direction to the top rail, does the top edge of the guardrail deflect to a height of 39 in (99.06 cm) or more above the walking or working level?
[29 CFR 1926.502(b)(4)] ☐ ☐ ☐

Note: For specifications on selection and construction of guardrail systems, refer to the OSHA guidelines in Appendix B of subpart M—Guardrail Systems.

4. Can midrails, screens, mesh, intermediate vertical members, solid panels, and equivalent structural members withstand (without failure) a force of at least 150 lb (68.039 kg) applied in any downward or outward direction at any point along the midrail or other member?
   [29 CFR 1926.502(b)(5)] ☐ ☐ ☐

5. Are guardrail systems surfaced to prevent snagging of clothing and injury from punctures or lacerations?
   [29 CFR 1926.502(b)(6)] ☐ ☐ ☐

6. Are workers PROHIBITED from allowing the ends of all top rails and midrails to overhang the terminal posts (unless overhang does not cause a hazard)?
   [29 CFR 1926.502(b)(7)] ☐ ☐ ☐

7. Are workers PROHIBITED from constructing top rails or midrails with steel banding and plastic banding?
   [29 CFR 1926.502(b)(8)] ☐ ☐ ☐

8. Are top rails and midrails of at least 0.25 in (6.35 mm) nominal diameter or thickness?
   [29 CFR 1926.502(b)(9)] ☐ ☐ ☐

Note: This is to prevent cuts and lacerations.

9. If wire rope is used for top rails, is it flagged at 6-ft (1.829-cm) intervals (or less) with high-visibility material?
   [29 CFR 1926.502(b)(9)] ☐ ☐ ☐

10. When guardrail systems are used to protect hoisting areas, is a chain, gate, or removable guardrail section placed across the access opening between guardrail sections when hoisting operations are not taking place?
    [29 CFR 1926.502(b)(10)] ☐ ☐ ☐

11. When guardrail systems are used at holes, are they erected on all unprotected sides or edges of the hole?
    [29 CFR 1926.502(b)(11)] ☐ ☐ ☐

12. When guardrail systems are placed around holes, do only two (or fewer) sides have removable guardrail sections to allow the passage of materials?
    [29 CFR 1926.502(b)(12)] ☐ ☐ ☐

13. When a hole is not in use, is it closed over with a cover, or is a guardrail system provided along all unprotected sides or edges?
    [29 CFR 1926.502(b)(12)] ☐ ☐ ☐

14. Are guardrail systems equipped with a gate (or offset so that a person cannot walk directly into the hole) when they are placed around holes that are used as points of access (such as ladderways)?
    [29 CFR 1926.502(b)(13)] ☐ ☐ ☐

15. Are guardrail systems used on ramps and runways erected along each unprotected side or edge?
    [29 CFR 1926.502(b)(14)] ☐ ☐ ☐

16. Is manila, plastic, or synthetic rope that is used for top rails or midrails inspected frequently to ensure that it continues to meet the strength requirements as indicated in questions 2, 3, and 4?
    [29 CFR 1926.502(b)(15)] ☐ ☐ ☐

17. Are guardrails systems, used on walking/working surfaces 6 ft (1.829 m) or more in height, to prevent falls from unprotected sides, edges, holes, remote excavations, and wall openings? ☐ ☐ ☐

18. Are the top edge heights of guardrail system members 42 in (1.067 m) plus or minus 3 in (7.62 cm), above the walking/working surface? ☐ ☐ ☐

19. Are wire rope guardrail systems more than 0.25 in (0.61 cm) in diameter or greater and flagged at 6 ft (1.83 m) intervals? ☐ ☐ ☐

20. Are canopies, screens, or toe boards installed to prevent falling objects? ☐ ☐ ☐

21. Are labeled covers secured over holes? ☐ ☐ ☐

FALL PROTECTION   YES   NO   N/A

1. DOES a competent person decide if fall protection is feasible when workers are assembling a scaffold or taking it apart? ☐ ☐ ☐

2. Have employees been properly trained in the □ □ □
following issues?
- Manufacturer's recommendations, restrictions, □ □ □
instructions, and warnings?
- Location of appropriate anchorage points and □ □ □
attachment techniques?
- Problems associated with elongation, deceleration □ □ □
distance, method of use, inspection, and storage
of all arrest systems?
3. Has the free-fall distance been considered, so that a □ □ □
worker will not strike a lower surface or object before
a fall is arrested?
4. Is all fall-arrest equipment free of potential damage □ □ □
from welding, chemical corrosion, or sandblasts?
5. Are lanyard lengths as short as necessary and in no □ □ □
case greater than 6 ft (1.8 m)?
6. Do lanyards have a shock-absorbing feature to limit □ □ □
the arresting forces to 500-600 lbs (227-272 kgs)?
7. DO workers on suspended scaffolds ALWAYS
- Wear safety belts attached to life lines? □ □ □
- Firmly anchor lifelines to an overhead structure □ □ □
and not to the scaffold?
8. Has the entire horizontal lifeline system been □ □ □
designed and approved by a competent person?
9. Are workers PROHIBITED from tying knots from □ □ □
the lanyard to the lifeline? (Mechanical rope grabs or
fall arresters MUST be used.)
10. DOES the vertical lifeline have a minimum breaking □ □ □
strength of 5000 lbs (2200 kgs)?
11. DOES the system provide fall protection as the worker □ □ □
connects to and releases from the lifeline?
12. Is the rope or cable free from signs of war or abrasion? □ □ □
13. Is the hardware riding on the horizontal lifeline □ □ □
made of steel? (Aluminum is NOT permitted because
it wars excessively.)
14. Have the anchorages to which the lifeline is attached □ □ □
been designed and evaluated specifically for a
horizontal lifeline?

15. DO workers have fall protection when working on a scaffold that is more than 10 ft (3.05 m) above a level? ☐ ☐ ☐
16. Is the fall protection system compatible with the lifeline on which it is to be installed or operated? ☐ ☐ ☐
17. DO workers use a harness, and NOT a body belt for personal fall protection? ☐ ☐ ☐
18. DO safety harnesses fit snugly around the chest and leg areas? (Women should use a harness with an 'X' configuration in the front; men should ensure the leg/buttock straps are secured so they will not slip if a sudden force is applied—thereby preventing injury to gender specific body parts) ☐ ☐ ☐
19. ENSURE safety harness D-ring is in the middle of the back where it easily can be reached to hook and unhook the lanyard without wearer performing a contortionist act that would cause a fall. ☐ ☐ ☐
20. Are all anchorage points for body harnesses located at shoulder height? ☐ ☐ ☐
21. Are all anchorage points stable, substantial, and sufficiently strong to withstand twice the potential impact energy of the free fall? ☐ ☐ ☐
22. DO most scaffolds have guard rails on all open sides and ends? ☐ ☐ ☐

Note: On supported scaffolds, and some other types, guardrails or personal fall protection is enough. On most suspension scaffolds, however, both are needed.

A guard rail is NOT needed on the working side of a supported scaffold, when the platform is less than 14 in (cm) from the work—8 in (cm) for plastering and lathing

The open side of an outrigger scaffold MUST NEVER be more than 3 ft (cm) from the face of the building.

23. On a supported scaffold:
    - Is the top rail 38 in (96.52 cm) to 45 in (114.3 cm) above the platform? ☐ ☐ ☐
    - Is the top rail, on a supported scaffold, strong enough to hold 200 lb (90.72 kg)? ☐ ☐ ☐
    - Is the top rail, on single-point and two-point suspension scaffolds, strong enough to support 100 lb (45.36 kg)? ☐ ☐ ☐

24. Is there a mid-rail about halfway between the platform and the top rail? ☐ ☐ ☐
25. Is the mid-rail capable of holding 150 lb (68.04 kg)? ☐ ☐ ☐

Note: A top-rail is needed when mesh, screens, or panels are needed—unless the mesh was designed and installed to meet guardrail requirements).

26. DOES a scaffold walkway have no more than a 9.5-in (34.13 cm) gap between planks and a guardrail? ☐ ☐ ☐
27. Are workers instructed to NOT let junk collect on the scaffold—eliminating the hazard to potentially cause a worker to trip and fall? ☐ ☐ ☐

8.9 Regulations

OSHA updated regulations address types of scaffolds—such as catenary scaffolds, step and trestle ladder scaffolds, and multilevel suspended scaffolds—that were not covered by OSHA's existing scaffold standards. The final rule allows employers greater flexibility in the use of fall protection systems to protect employees working on scaffolds and to extend fall protection to erectors and dismantlers of scaffolds to the feasible extent.

29 CFR 1910.28

.28(b)	Wood pole scaffolds		.28(k)	Carpenter's bracket scaffolds
.28(c)	Tube and coupler scaffolds		.28(l)	Bricklayer's square scaffolds
.28(d)	Tubular welded frame scaffolds		.28(m)	Horse scaffolds
.28(e)	Outrigger scaffold		.28(n)	Needle-beam scaffolds
.28(f)	Mason's adjustable, multiple-point suspension scaffolds		.28(o)	Plasterers', decorators', and large-area scaffolds
.28(g)	Two-point suspension scaffolds (swinging scaffolds)		.28(p)	Interior-hung scaffolds
			.28(q)	Ladder-jack scaffolds
			.28(r)	Window-jack scaffolds
.28(h)	Stone setters' adjustable, multiple-point suspension scaffolds		.28(s)	Roofing brackets
			.28(t)	Crawling board or chicken ladders
.28(i)	Single-point adjustable suspension scaffolds		.28(u)	Float or ship scaffolds
.28(j)	Boatswain's chair			

29 CFR.1926.450 Subpart L—Scaffolds

.451(a)	Capacity	.452(k)	Ladder jack scaffolds
.451(b)	Scaffold platform construction	.452(l)	Window jack scaffolds
.451(c)	Criteria for supported scaffolds	.452(m)	Crawling boards (chicken ladders)
.451(d)	Criteria for suspension scaffolds	.452(n)	Step, platform, and trestle ladder scaffolds
.451(e)	Access	.452(o)	Single-point adjustable suspension scaffolds
.451(f)	Use	.452(p)	Two-point adjustable suspension scaffolds (swing stages)
.451(g)	Fall protection		
.451(h)	Falling object protection	.452(q)	Multi-point adjustable suspension scaffolds
.452(a)	Pole scaffolds	.452(r)	Catenary scaffolds
.452(b)	Tube and coupler scaffolds	.452(s)	Float (ship) scaffolds
.452(c)	Fabricated frame scaffolds (tubular welded frame scaffolds)	.452(t)	Interior hung scaffolds
		.452(u)	Needle beam scaffolds
.452(d)	Plasterers', decorators', and large-area scaffolds	.452(v)	Multilevel suspended scaffolds
		.452(w)	Mobile scaffolds
.452(e)	Bricklayers' square scaffolds (squares)	.452(x)	Repair bracket scaffolds
.452(f)	Horse scaffolds	.453(y)	Aerial lifts manlifts, and scissor lifts
.452(g)	Form scaffolds and carpenters' bracket scaffolds	App. A	Scaffold specifications
		App. B	Criteria for determining the feasibility of providing safe access and fall protection for scaffold erectors and dismantlers
.452(h)	Roof bracket scaffolds		
.452(i)	Outrigger scaffolds		
.452(j)	Pump jack scaffolds		

8.10 Standards (Consensus)

ANSI American National Standards Institute

ASC (Accredited Standards Committee) A92—Aerial Platforms Committee—has developed six current American National Standards—safety guides for the aerial platform industry:

    A92.2   Vehicle-mounted rotating and elevating work platforms
    A92.3   Manually propelled elevating work platforms

A92.5 Boom-supported elevating work platforms
A92.6 Self-propelled elevating work platforms
A92.7 Airline ground support vehicle-mounted vertical lift devices
A92.8 Vehicle-mounted bridge inspection and maintenance devices
A92.9 Mast-climbing work platforms
A92.10 Transport platforms

ANSI/SIA—Scaffolds and Other Elevating Devices
A92 Elevating and Vehicle Lift Devices Package
A92.2 2009 Vehicle-Mounted Elevating and Rotating Aerial Devices
A92.3-2006 Propelled Elevating Aerial Platforms—American National Standard
A92.5-2006 Boom-Supported Elevating Work Platforms—American National Standard
A92.6-2006 Self-Propelled Elevating Work Platforms
A92.8-2006 Vehicle-Mounted Bridge Inspection and Maintenance Devices
A92.10-2009 Transport Platforms

SIA—Scaffold Industry Association

Maintains active representation on various committees accredited by ANSI that develops scaffold standards affecting SIA members:

A10.8 Safety requirements for scaffolding
A92.2 Vehicle-mounted elevating and rotating aerial devices
A92.3 Manually propelled elevating work platforms
A92.5 Boom-supported elevating work platforms
A92.6 Self-propelled elevating work platforms
A92.7 Airline ground support vehicle-mounted vertical lift devices
A92.8 Vehicle-mounted bridge inspection and maintenance devices
A92.9 Climbing work platforms
A92.10 Transport platforms
A120.1 Powered platforms for building maintenance
Z359 Standards for fall protection

Chapter 9

Fall Protection

OSHA regulations require INSPECTION of personal fall arrest systems (PFAS) prior to each use for mildew, wear, damage, and other deterioration, and the immediate REMOVAL of defective components from service if their strength or function may be adversely affected.

Check for OSHA-recognized test methods for evaluating the performance of fall arrest systems, including Strength Test, Force Test, and Deceleration Device Test. Not all systems may need to be individually tested; however, the performance of some systems may be based on data and calculations derived from testing of similar systems, provided that enough information is available to demonstrate similarity of function and design.

Ideally, a personal fall arrest system is designed, tested, and supplied as a complete system. However, it is common practice for lanyards, connectors, lifelines, deceleration devices, body belts, and body harnesses to be interchanged since some components wear out before others.

The employer and employee should realize that not all components are interchangeable. For instance, a lanyard should not be connected between a body belt (or harness) and a deceleration device of the self-retracting type since this can result in additional free fall for which the system was not designed.

Any substitution or change to a personal fall arrest system should be fully evaluated or tested by a competent person to determine that it meets the standard, before the modified system is put in use.

9.1 Fall Arrest Systems

Harnesses, retractable lanyards, and static lines are effective in minimizing the risk of working at height. To do this properly, however, they must be in good working order—therefore this equipment must be used, stored, inspected, and cleaned correctly.

To maintain service life and high performance, all belts and harnesses MUST be inspected frequently. Visual inspection before each use should become routine, but a more thorough inspection by a trained and competent person MUST be done at regular intervals.

- Inspection only includes
  - Visual Inspection of item
  - Tagging of item after inspection
  - Entering log record
- Cleaning includes
  - Washing and drying of all synthetic materials
  - Cleaning of all metal components

Maintenance guidelines

A sound fall protection plan MUST include proper care and maintenance of all fall protection equipment along with the associated items necessary for a complete system.

- CONDUCT training sessions, related to care and maintenance, at regular intervals depending on the nature of the work.

Note: ANSI and OSHA standards require that training be performed by a competent person. The user is responsible for ensuring that he or she knows how to properly inspect, use, store, and maintain the equipment.

- Store personal protection equipment (harnesses, lanyards, etc.) along with connectors and other related items in a clean, dry environment free from direct sunlight, dust, excessive heat, and harmful chemicals.
- Clean personal protection equipment periodically using a mild detergent and water.
- Wash with a soft, nonabrasive brush or sponge and allow to air-dry after removing the excess water with a dry cloth.
- DO NOT put personal protection equipment in clothes dryer or use blow dryer; excessive heat may melt the webbing and alter the strength.
- DO NOT use chemicals to clean heavily soiled gear; chemicals may destroy the webbing.
- Ensure that each user inspects the fall arrest system before each use.

Note: User should check before each use to be sure a formal inspection has been performed within the last 6 months.

- Remove, immediately from service, any harness having excessive wear and aging (generally not repairable).
- Dispose of harness to a competent person authorized to perform inspections.
- Ensure that a competent person inspects all personal fall arrest systems at intervals of no more than 6 months.
  - Harnesses, lanyards, and synthetic lifelines manufactured from webbing or rope, or both (products used in typical fall arrest system)
  - Webbing (straps) and stitching for cuts, fraying, pulled or broken threads, abrasion, excessive wear, altered or missing straps, burn, heat, and chemical exposures
  - All rope for cuts, fraying, pulled or broken strands, abrasion, excessive wear, burns, and heat and chemical exposure
  - All metallic parts (such as D-rings, snap hooks, buckles, adjusters, and grommets) for deformation, fractures, cracks, corrosion, deep pitting, sharp edges, cuts, deep nicks, missing or loose parts, improper function, evidence of burns, excessive heat, and chemical exposure
- Perform more frequent formal inspections if a harness is exposed to sever working conditions.
- Establish frequency of inspection by a competent person based on such factors as the nature and severity of the workplace conditions and exposure time of the equipment.

Note: Verify that the competent person performs the inspection following the procedures previously outlined.

- Record results in the formal inspection log.
- MAKE SURE the user checks the log before each use to be sure a formal inspection has been performed within the last 6 months.

Note: Harnesses MUST be immediately removed from service, when user inspection, in accordance with stated inspection procedures, reveals signs of inadequate maintenance

- Establish and enforce a policy and procedure to immediately remove from service any personal fall arrest system (or parts thereof) that are found to be defective, damaged, impact loaded, or in need of maintenance.

- DO NOT make any repairs to defective or damaged fall systems; only the manufacturer should repair fall protection equipment and systems.

Source: French Creek Production, Inc., Franklin, PA (fcp@velocity.net)

Regulations

This checklist covers regulations issued by the U.S. Department of Labor, Occupational Safety and Health Administration (OSHA) under the general industry standards 29 CFR 1910.66, and the construction standards 29 CFR 1926.501 to 1926.503.

It applies to temporary worksites associated with construction, alteration, demolition, and repair work, including painting and decorating.

The regulations cited apply only to private employers and their employees, unless adopted by a state agency and applied to other groups such as public employees. They DO NOT address safety net systems or positioning device systems. (Consult OSHA regulations.)

FALL SYSTEM COMPONENTS   YES   NO   N/A

1. Are connectors made of drop-forged, pressed or formed steel, or of equivalent materials? ☐ ☐ ☐

2. Do connectors have a corrosion-resistant finish? ☐ ☐ ☐

3. Are all surfaces and edges smooth to prevent damage to interfacing parts of the system?. ☐ ☐ ☐

4. Do lanyards and vertical lifelines, which tie-off one employee, have a minimum breaking strength of 5000 lb (2267.96 kg)? ☐ ☐ ☐

5. Do self-retracting lifelines and lanyards, which automatically limit free fall distance to 2 ft (0.61 m) or less, have components capable of sustaining a minimum static tensile load of 3000 lb (1360.78 kg) applied to the device with the lifeline or lanyard in the fully extended position? ☐ ☐ ☐

6. Are self-retracting lifelines and lanyards, which do not limit free fall distance to 2 ft (0.61 m) or less, rip stitch lanyards, and tearing and deforming lanyards capable of sustaining a minimum tensile load of 5000 lb (2267.96 kg) applied to the device with the lifeline or lanyard in the fully extended position? ☐ ☐ ☐

7. Are D-rings and snap hooks capable of sustaining a minimum tensile load of 5000 lb (2267.96 kg)? ☐ ☐ ☐

Fall Protection

8. Are D-rings and snap hooks 100 percent proof-tested to a minimum tensile load of 3600 lb (1632.93 kg) without cracking, breaking, or taking permanent deformation? ☐ ☐ ☐

9. Are snap hooks sized to be compatible with the member to which they are connected so as to prevent unintentional disengagement of the snap hook by depression of the snap hook keeper by the connected member? ☐ ☐ ☐
   - Or, are they locking-type snap hooks, designed and used to prevent disengagement of the snap hook by the contact of the snap hook keeper by the connected member? ☐ ☐ ☐

10. Are horizontal lifelines, where used, designed and installed as part of complete personal fall arrest system, which maintains a safety factor of at least 2, under the supervision of a qualified person? ☐ ☐ ☐

11. Are anchorages, to which personal fall arrest equipment is attached, capable of supporting at least 5000 lb (2267.96 kg) per employee attached? ☐ ☐ ☐
    - Or, are they designed, installed, and used as part of complete personal fall arrest system that maintains a safety factor of at least 2, under supervision of a qualified person? ☐ ☐ ☐

12. Are ropes and straps (webbing), used in lanyards, lifelines, and strength components of body belts and body harnesses, made from synthetic fibers or wire rope? ☐ ☐ ☐

System Performance Criteria

13. Do personal fall arrest systems, when stopping a fall,
    - Limit maximum arresting force on an employee to 900 lb (408.23 kg) when used with a body belt? ☐ ☐ ☐
    - Limit maximum arresting force on an employee to 1800 lb (816.47 kg) when used with a body harness? ☐ ☐ ☐
    - Bring an employee to a complete stop and limit maximum deceleration distance an employee travels to 3.5 ft (1.07 m)? ☐ ☐ ☐

200 Inspections

- Have sufficient strength to withstand twice the potential impact energy of an employee free falling a distance of 6 ft (1.8 m), or the free fall distance permitted by the system, whichever is less? ☐ ☐ ☐

Note:
- When used by employees having a combined person and tool weight of less than 310 lb (140 kg) personal fall arrest systems which meet the criteria and protocols contained in the preceding paragraphs shall be considered as complying with the provisions of preceding paragraphs
- When used by employees having a combined tool and body weight of 310 lb (140 kg) or more, personal fall arrest systems which meet the criteria and protocols contained in the preceding paragraphs may be considered as complying with the provisions of the preceding paragraphs, provided that the criteria and protocols are modified appropriately to provide proper protection for such heavier weights.

Care and Use

14. Is engaging a snap hook PROHIBITED, unless the snap hook is of a locking type designed and used to prevent disengagement from the following direct connections:
    - To webbing, rope or wire rope? ☐ ☐ ☐
    - To each other? ☐ ☐ ☐
    - To a D-ring to which another snap hook or other connector is attached? ☐ ☐ ☐
    - To a horizontal lifeline? ☐ ☐ ☐
    - To any object which is incompatibly shaped or dimensioned in relation to the snap hook such that the connected object could depress the snap hook keeper a sufficient amount to release itself? ☐ ☐ ☐

15. Are devices used to connect to a horizontal lifeline, which may become a vertical lifeline, capable of locking in either direction on the lifeline? ☐ ☐ ☐

16. Are personal fall arrest systems rigged such that a worker can neither free-fall more than 5 ft (1.8 m), nor contact any lower level? ☐ ☐ ☐

17. Is the attachment point of the body belt located in the center of the wearer's back? ☐ ☐ ☐

18. Is the attachment point of the body harness located in the center of the wearer's back near shoulder level or above the wearer's head? ☐ ☐ ☐

19. Is each worker provided a separate lifeline, when vertical lifelines are used? ☐ ☐ ☐
20. Are personal fall arrest systems or components used only for employee fall protection? ☐ ☐ ☐
21. Are personal fall arrest systems or components subjected to impact loading immediately removed from service and NOT be used again for employee protection unless inspected and determined by a competent person to be undamaged and suitable for reuse? ☐ ☐ ☐
22. The employer shall provide for prompt rescue of employees in the event of a fall or shall ensure the self-rescue capability of employees? ☐ ☐ ☐
23. Before using a personal fall arrest system and after any component or system is changed, employees shall be trained in accordance with the requirements of paragraph 1910.66(i)(1), in the safe use of the system? ☐ ☐ ☐

Source: OSHA Occupational Safety and Health Administration.

PERSONAL FALL ARREST SYSTEMS   YES   NO   N/A

1. Are workers on scaffolds that are more than 10 ft (3.05 m) above a lower level protected from falling by one of the following measures:
   - Personal fall arrest system for workers on ladder jack scaffolds? ☐ ☐ ☐
     [29 CFR 1926.451(g)(1)(i)]
   - Guardrail system installed within 9.5 in (24.13 cm) of and along at least one side of the walkway for workers on a walkway located within a scaffold? ☐ ☐ ☐
     [29CFR 1926.451(g)(1)(v)]
   - Personal fall arrest system or guardrail system to protect workers doing overhand bricklaying from a supported scaffold from falling off open sides and ends of the scaffold (except at the side next to the wall being laid)? ☐ ☐ ☐
     [29 CFR 1926.451(g)(1)(vi)]
   - Personal fall arrest system or guardrail system for workers on all other scaffolds? ☐ ☐ ☐
     [29 CFR 1926.451(g)(1)(vii)]

Inspections

2. Does a competent person determine the feasibility and safety of providing fall protection for workers erecting or dismantling supported scaffolds?
[29 CFR 1926.451(g)(2)] ☐ ☐ ☐

3. Are personal fall arrest systems, designed and installed by a qualified person, to maintain a safety factor of at least 2? ☐ ☐ ☐

4. Do personal fall arrest systems, and their use, comply with the provisions set forth in [29 CFR 1926.502(d)(1)], including
   - Are workers PROHIBITED from using body belts as part of a personal fall arrest system? ☐ ☐ ☐

Note: The use of a body belt in a positioning device system is acceptable and is regulated under [29 CFR 1926.502 (e)].

   - Are fall system connectors drop-forged, pressed or formed steel, or made of equivalent materials? [29 CFR 1926.502(d)(2)] ☐ ☐ ☐

5. Do workers erecting or dismantling supported scaffolds use fall protection when it is safe and feasible?
[29 CFR 1926.451(g)(2)] ☐ ☐ ☐

6. Are personal fall arrest systems used on scaffolds attached by a lanyard to a vertical lifeline, horizontal lifeline, or scaffold structural member?
[29 CFR 1926.451(g)(3)] ☐ ☐ ☐

7. When vertical lifelines are used, are they fastened to a fixed safe point of anchorage, independent of the scaffold, and protected from sharp edges and abrasion?
[29 CFR 1926.451(g)(3)(i)] ☐ ☐ ☐

Note: Safe points of anchorage include structural members of buildings, but do not include standpipes, vents, other piping systems, electrical conduit, outrigger beams, or counterweights.

8. When horizontal lifelines are used, are they secured to two or more structural members of the scaffold?
[29 CFR 1926.451(g)(3)(ii)] ☐ ☐ ☐

9. Are workers PROHIBITED from attaching vertical lifelines and independent support lines to one another, to the same point of anchorage, and to the same point on the scaffold or personal fall arrest system?
[29 CFR 1926.451(g)(3)(iv)] ☐ ☐ ☐

Fall Protection

10. When guardrail systems are required, are they installed along all open sides and ends of platforms?
    [29 CFR 1926.451(g)(4)(i)]

    ☐ ☐ ☐

Note: Guardrails systems must be installed before the scaffold is used by workers other than erecting or dismantling crews.

11. If the scaffolds were manufactured or placed in service after January 1, 2000, is the top edge height of top rails (or equivalent member) on supported scaffolds between 38 in (0.97m) and 45 in (1.15 m) above the platform surface?
    [29 CFR 1926.451(g)(4)(ii)]

    ☐ ☐ ☐

Note: When necessary, the height of the top edge may exceed 45 in (1.15 m) if the guardrail meets all OSHA requirements.

12. Are midrails, screens, mesh, intermediate vertical members, and solid panels (or equivalent structural members) installed between the top edge of the guardrail system and the scaffold platform?
    [29 CFR 1926.451(g)(4)(iii)].

    ☐ ☐ ☐

13. When midrails are used, are they installed approximately midway between the top edge of the guardrail system and the platform surface?
    [29 CFR 1926.451(g)(4)(iv)]

    ☐ ☐ ☐

14. When screens and mesh are used, do they extend from the top edge of the guardrail system to the scaffold platform, and along the entire opening between the supports?
    [29 CFR 1926.451(g)(4)(v)]

    ☐ ☐ ☐

15. When intermediate members (such as balusters or additional rails) are used, are they installed 19 in or less apart?
    [29 CFR 1926.451(g)(4)(vi)]

    ☐ ☐ ☐

16. Can each top rail (or equivalent member) of a guardrail system withstand (without failure) a 200-lb (90.72-kg) force applied in any downward or horizontal direction at any point along its top edge?
    [29 CFR 1926.451(g)(4)(vii)]

    ☐ ☐ ☐

Note: Appendix A of subpart L of the OSHA regulations gives directions for constructing acceptable guardrail systems.

17. When a 200-lb (90.72-kg) force is applied in a downward direction on the top rail (or equivalent member) of a guardrail system, does the top edge still maintain the OSHA-required height?
[29 CFR 1926.451(g)(4)(viii)] ☐ ☐ ☐

18. Can midrails, screens, mesh, intermediate vertical members, solid panels, and equivalent structural members of a guardrail system withstand (without failure) a 150-lb (68.04-kg) force applied in any downward or horizontal direction at any point along the midrail or other member?
[29 CFR 1926.451(g)(4)(ix)] ☐ ☐ ☐

19. Are guardrails surfaced to prevent snagging of clothing and injury from punctures or lacerations?
[29 CFR 1926.451(g)(4)(xi)] ☐ ☐ ☐

20. Is it prohibited for rails to overhang the terminal posts, except when such overhang does not constitute a projection hazard?
[29 CFR 1926.451(g)(4)(xii)] ☐ ☐ ☐

21. Is the use of steel or plastic banding for top rails or midrails prohibited?
[29 CFR 1926.451(g)(4)(xiii)] ☐ ☐ ☐

22. If manila, plastic, or other synthetic rope is used for top rails or midrails, is it inspected by a competent person as necessary to ensure that it continues to meet the OSHA strength requirements?
[29 CFR 1926.451(g)(4)(xiv)] ☐ ☐ ☐

23. If crossbracing is used to replace a midrail, is the crossing point of the two braces between 20 in (50.8 cm) and 30 in (76.2 cm) above the work platform?
[29 CFR 1926.451(g)(4)(xv)] ☐ ☐ ☐

24. If crossbracing is used to replace a top rail, is the crossing point of the two braces between 38 in (96.52 cm) and 48 in (1.22 m) above the work platform?
[29 CFR 1926.451(g)(4)(xv)] ☐ ☐ ☐

25. If crossbracing is used to replace a midrail or top rail, are the end points at each upright 48 in (1.22 m) apart or less?
[29 CFR 1926.451(g)(4)(xv)] ☐ ☐ ☐

FALLING OBJECT PROTECTION

	YES	NO	N/A

1. Do workers on scaffolds wear hardhats? ☐ ☐ ☐
[29 CFR 1926.451(h)(1)]

2. Are workers protected from falling hand tools, debris, ☐ ☐ ☐
and other small objects by toe boards, screens,
guardrail systems, debris nets, catch platforms, or
canopy structures that contain or deflect the falling
objects?
[29 CFR 1926.451(h)(1)]

3. If objects are too large, heavy, or massive to be ☐ ☐ ☐
contained or deflected, are they moved away from
the edge of the surface from which they could fall
and secured?
[29 CFR 1926.451(h)(1)]

4. If tools, materials, or equipment could fall from a
scaffold and strike workers, is one of the following
protective measures taken:
   - Barricade the area below the scaffold, to which ☐ ☐ ☐
objects can fall?
   - PROHIBIT workers from entering the hazard ☐ ☐ ☐
area?
   - Keep objects that may fall far enough away from ☐ ☐ ☐
the edge of a higher level so that those objects
would not go over the edge if they were displaced?
   - Erect toe board along the edge of platforms more ☐ ☐ ☐
than 10 ft (3.05 cm) above lower levels for a
distance sufficient to protect workers below?
   - Keep potential fall objects far enough away from ☐ ☐ ☐
the edge of the higher level so that those objects
would not go over the edge if they were displaced?
   - Erect a canopy structure, debris net, or catch ☐ ☐ ☐
platform?
[29 CFR 1926.501(c)]

5. If tools, materials, or equipment are piled higher than
the top edge of the toe board, is one of the following
protective measures taken:
   - Erect paneling or screening, extending from the ☐ ☐ ☐
toe board or platform to the top of the guardrail,
for a distance sufficient to protect the workers
below?

- Install guardrail system having openings small enough to prevent passage of falling objects? ☐ ☐ ☐
- Erect canopy structure, debris net, or catch platform erected over the workers that is strong enough to withstand the impact forces of the falling objects? ☐ ☐ ☐
[29 CFR 1926.451(h)(2)]

6. If canopies are used to protect workers, are they installed between the falling object hazard and the workers? ☐ ☐ ☐
[29 CFR 1926.451(h)(3)(i)]

7. If toe boards are used to protect workers, can they withstand (without failure) a force of at least 50 lb (22.67 kg) applied in any downward or horizontal direction at any point along the toe board? ☐ ☐ ☐
[29 CFR 1926.451(h)(4)(i)]

Note: Appendix A of subpart L of the OSHA regulations provides directions for constructing acceptable toe boards.

8. If toe boards are used to protect workers, are they at least 3.5 in (88.9 mm) high from the top edge of the toe board to the level of the walking/working surface? ☐ ☐ ☐
[29 CFR 1926.451(h)(4)(ii)]

9. If toe boards are used to protect workers, are they securely fastened in place at the outermost edge of the platform? ☐ ☐ ☐
   - Do they have 0.25-in (6.35-mm) or less clearance above the walking or working surface? ☐ ☐ ☐
   [29 CFR 1926.451(h)(4)(ii)]

10. If toe boards are used to protect workers, are they solid or with openings of 1 in (25.4 mm) or less in the greatest dimension? ☐ ☐ ☐
[29 CFR 1926.451(h)(4)(ii)]

WALKING/WORKING SURFACES   YES   NO   N/A

1. Do walking and working surfaces have the strength and structural integrity to support people safely? ☐ ☐ ☐
[29 CFR 1926.501(a)(2)]

2. Are employees PROHIBITED from working on walking and working surfaces that cannot support them safely?
[29 CFR 1926.501(a)(2)]   ☐ ☐ ☐

3. If fall protection systems are required, are they installed before employees begin work?   ☐ ☐ ☐

Note: This requirement is not found in 1926 .502(a)(2).

Unprotected Sides and Edges

4. Do guardrail systems, safety net systems, or personal fall arrest systems protect employees when they work on unprotected sides and edges of walking and working surfaces that are 6 ft (1.83 m) or more above a lower level?
[29 CFR 1926.501(b)(1)]   ☐ ☐ ☐

5. Do guardrail systems, safety net systems, or personal fall arrest systems protect employees during construction of leading edges 6 ft (1.83 m) or more above lower levels?
[29 CFR 1926.501(b)(2)(i)]   ☐ ☐ ☐

Note: Exceptions are permitted if these systems are infeasible or create a greater hazard. However, a fall protection plan must still be developed and implemented.

Hoist Area

6. Do guardrail systems or personal fall arrest systems protect employees in a hoist area from falling 6 ft (1.83 m) or more to lower levels?
[29 CFR 1926.501(b)(3)]   ☐ ☐ ☐

7. Does a personal fall arrest system protect employees if guardrail systems are removed for hoisting operations, requiring employees to lean through the access opening or out over the edge of the access opening (e.g., to receive or guide equipment and materials)?
[29 CFR 1926.501(b)(3)]   ☐ ☐ ☐

Holes

8. Do personal fall arrest systems, covers, or guardrail systems erected around holes protect employees on walking and working surfaces more than 6 ft (1.83 m) above lower levels from falling through holes (including skylights)?
[29 CFR 1926.501(b)(4)(i)] ☐ ☐ ☐

9. Do covers protect employees on a walking and working surface from tripping in or stepping into holes (including skylights)?
[29 CFR 1926.501(b)(4)(ii)] ☐ ☐ ☐

10. Do covers protect employees on a walking and working surface from objects falling through holes (including skylights)?
[29 CFR 1926.501(b)(4)(iii)] ☐ ☐ ☐

Formwork and Reinforcing Steel

11. Do personal fall arrest systems, safety net systems, or positioning device systems protect employees on the face of formwork or reinforcing steel from falling 6 ft (1.83 m) or more to lower levels?
[29 CFR 1926.501(b)(5)] ☐ ☐ ☐

Ramps, Runways, and Other Walkways

12. Do guardrail systems protect employees on ramps, runways, and other walkways from falling 6 ft (1.83 m) or more to lower levels?
[29 CFR 1926.501(b)(6)] ☐ ☐ ☐

Excavations

13. Do guardrail systems, fences, or barricades protect employees from falling at the edge of an excavation 6 ft (1.83 m) or more in depth when the excavation is blocked because of plant growth or other visual barrier?
[29 CFR 1926.501(b)(7)(i)] ☐ ☐ ☐

14. Do guardrail systems, fences, barricades, or covers protect employees from falling at the edge of a well, pit, shaft, and similar excavation 6 ft (1.83 m) or more in depth?
[29 CFR 1926.501(b)(7)(ii)] ☐ ☐ ☐

Dangerous Equipment

15. Do guardrail systems or equipment guards protect employees from falling from less than 6 ft (1.83 m) onto dangerous equipment?
    [29 CFR 1926.501(b)(8)(i)]   ☐ ☐ ☐

16. Do guardrail systems, personal fall arrest systems, or safety net systems protect employees 6 ft (1.83 m) or more above dangerous equipment from fall hazards?
    [29 CFR 1926.501(b)(8)(ii)]   ☐ ☐ ☐

Overhead Bricklaying and Related Work

17. Do guardrail systems, safety net systems, or personal fall arrest systems protect employees from falling while they perform overhand bricklaying and related work 6 ft (1.83 m) or more above lower levels? Or, are employees restricted to working in a controlled-access zone?
    [29 CFR 1926.501(b)(9)(i)]   ☐ ☐ ☐

Note: Exceptions are permitted if these systems are infeasible or create a greater hazard; however, a fall protection plan must still be developed and implemented.

18. Does a guardrail system, safety net system, or personal fall arrest system protect employees reaching more than 10 in (15.4 cm) below the level of the walking and working surface?
    [29 CFR 1926.501(b)(9)(ii)]   ☐ ☐ ☐

Roofing Work on Low-Slope Roofs

19. Does one of the following systems protect employees from falling while they work on low-slope roofs with unprotected sides and edges 6 ft (1.83 m) or more above lower levels:
    - Guardrail systems?   ☐ ☐ ☐
    - Safety net systems?   ☐ ☐ ☐
    - Personal fall arrest systems?   ☐ ☐ ☐
    - Combination warning line system and safety net system?   ☐ ☐ ☐

- Warning line system and personal fall arrest system?   ☐ ☐ ☐
- Warning line system and safety monitoring system?   ☐ ☐ ☐
[29 CFR 1926.501(b)(10)]

Note: Exceptions are permitted if these systems are infeasible or create a greater hazard; however, a fall protection plan must still be developed and implemented. On roofs 50 ft (15.24 m) or less in width, using a safety-monitoring system alone (i.e., without the warning line system) is also permitted. (See Appendix A to subpart M—Determining Roof Widths of 29 CFR 1926 for help.)

Steep Roofs

20. Do guardrail systems with toe boards, safety net systems, or personal fall arrest systems protect employees from falls off a steep roof with unprotected sides and edges 6 ft (1.83 m) or more above lower levels?   ☐ ☐ ☐
[29 CFR 1926.501(b)(11)]

Precast Concrete Erection

21. Do guardrail systems, safety net systems, or personal fall arrest systems protect employees 6 ft (1.83 m) or more above lower levels when they are engaged in erecting precast concrete members and related operations?   ☐ ☐ ☐
[29 CFR 1926.501(b)(12)]

Note: Exceptions are permitted if these systems are infeasible or create a greater hazard; however, a fall protection plan MUST still be developed and implemented.

Wall Openings

22. Are employees protected from falling by a guardrail system, a safety net system, or a personal fall arrest system if they are working on, at, above, or near wall openings (including those with chutes attached) where
    - Outside bottom edge of the wall opening is 6 ft (1.83 m) or more above lower levels?   ☐ ☐ ☐
    - Inside bottom edge of the wall opening is less than 39 in (99.06 cm) above the walking and working surface?   ☐ ☐ ☐
[29 CFR 1926.501(b)(14)]

Fall Protection

Walking and Working Surfaces (Not Otherwise Addressed)

23. Does a guardrail system, safety net system, or personal fall arrest system protect employees on a walking or working surface 6 ft (1.83 m) or more above lower levels that have not been addressed as part of this checklist?
[29 CFR 1926.501(b)(15)]   ☐ ☐ ☐

GUARDRAIL SYSTEMS   YES   NO   N/A

1. Is the top edge height of top rails (or equivalent guardrail system members) 42 in (1.143 m), plus or minus 3 in (7.62 cm) above the walking or working level?
[29 CFR 1926.502(b)(1)]   ☐ ☐ ☐

Note: When necessary, the height of the top edge may exceed the 45 in (1.067 cm), if the guardrail system meets all other criteria. When employees are using stilts, the top edge height of the top rail (or equivalent member) shall be increased by an amount equal to the height of the stilts.

2. When no wall or parapet wall is at least 21 in (53.34 cm) high, are the midrails, screens, mesh, or intermediate vertical members (or equivalent intermediate structural members) installed between the top edge of the guardrail system and the walking or working surface?
[29 CFR 1926.502(b)(2)]   ☐ ☐ ☐

3. Are midrails installed at a height midway between the top edge of the guardrail system and the walking or working level?
[29 CFR 1926.502(b)(2)(i)]   ☐ ☐ ☐

4. Do screens and mesh extend from the top rail to the walking or working level and along the entire opening between top rail supports?
[29 CFR 1926.502(b)(2)(ii)]   ☐ ☐ ☐

5. When used between posts, are intermediate members (such as balusters) 19 in (48.26 cm) apart or less?
[29 CFR 1926.502(b)(2)(iii)]   ☐ ☐ ☐

6. Are other structural members (such as additional midrails and architectural panels) installed so that openings in the guardrail system are 19 in (48.26 cm) wide or less?
[29 CFR 1926.502(b)(2)(iv)]   ☐ ☐ ☐

7. Can guardrail systems withstand (without failure) a force of at least 200 lb (90.72 kg) applied within 2 in (5.08 cm) of the top edge, in any outward or downward direction, at any point along the top edge?
[29 CFR 1926.502(b)(3)]   ☐ ☐ ☐

8. When a 200 lb (90.72 kg) test load is applied in a downward direction to the top rail, does the top edge of the guardrail deflect to a height of 39 in (99.06 cm) or more above the walking or working level?
[29 CFR 1926.502(b)(4)]   ☐ ☐ ☐

Note: For specifications on selection and construction of guardrail systems, please refer to the OSHA guidelines in Appendix B of subpart M—Guardrail Systems.

9. Can midrails, screens, mesh, intermediate vertical members, solid panels, and equivalent structural members withstand (without failure) a force of at least 150 lb (68.04 kg) applied in any downward or outward direction at any point along the midrail or other member?
[29 CFR 1926.502(b)(5)]   ☐ ☐ ☐

10. Are guardrail systems surfaced to prevent snagging of clothing and injury from punctures or lacerations?
[29 CFR 1926.502(b)(6)]   ☐ ☐ ☐

11. Are workers PROHIBITED from allowing the ends of all top rails and midrails to overhang the terminal posts (unless overhang does not cause a hazard)?
[29 CFR 1926.502(b)(7)]   ☐ ☐ ☐

12. Are workers PROHIBITED from constructing top rails or midrails of steel banding and plastic banding?
[29 CFR 1926.502(b)(8)]   ☐ ☐ ☐

13. Are top rails and midrails of at least 0.25 in nominal diameter or thickness?
[29 CFR 1926.502(b)(9)]   ☐ ☐ ☐

Note: This is to prevent cuts and lacerations.

14. If wire rope is used for top rails, is it flagged at 6-ft (1.83-m) intervals (or less) with high-visibility material?
[29 CFR 1926.502(b)(9)]   ☐ ☐ ☐

Fall Protection

15. When guardrail systems are used to protect hoisting areas, is a chain, gate, or removable guardrail section placed across the access opening between guardrail sections when hoisting operations are not taking place?
[29 CFR 1926.502(b)(10)] ☐ ☐ ☐

16. When guardrail systems are used at holes, are they erected on all unprotected sides or edges of the hole?
[29 CFR 1926.502(b)(11)] ☐ ☐ ☐

17. When guardrail systems are placed around holes, do only two (or fewer) sides have removable guardrail sections to allow the passage of materials?
[29 CFR 1926.502(b)(12)] ☐ ☐ ☐

18. When a hole is not in use, is it closed over with a cover, or is a guardrail system provided along all unprotected sides or edges?
[29 CFR 1926.502(b)(12)] ☐ ☐ ☐

19. Are guardrail systems equipped with a gate (or offset so that a person cannot walk directly into the hole) when they are placed around holes that are used as points of access (such as ladderways)?
[29 CFR 1926.502(b)(13)] ☐ ☐ ☐

20. Are guardrail systems used on ramps and runways erected along each unprotected side or edge?
[29 CFR 1926.502(b)(14)] ☐ ☐ ☐

21. Is manila, plastic, or synthetic rope that is used for top rails or midrails inspected frequently to ensure that it continues to meet the strength requirements indicated in questions 7, 8, and 9?
[29 CFR 1926.502(b)(15)] ☐ ☐ ☐

TRAINING   YES NO N/A

1. Has a training program been provided to everyone who might be exposed to fall hazards?
[29 CFR 1926.503(a)(1)] ☐ ☐ ☐

Note: The training program must enable each employee to recognize the hazards of falling and know the procedures for minimizing these hazards.

2. Is the training program conducted by a competent person?
[29 CFR 1926.503(a)(2)]   ☐ ☐ ☐

3. Have individual certification records been prepared that contain the name or other identity of the person trained, the date(s) of the training, and the signature of the employer or person who conducted the training?
[29 CFR 1926.503(b)(1)]   ☐ ☐ ☐

4. Is the latest training certification for all persons trained available for inspection?
[29 CFR 1926.503(b)(2)]   ☐ ☐ ☐

5. Is retraining conducted if a person can no longer recognize the hazards of falling or follow the proper procedures?
[29 CFR 1926.503(c)]   ☐ ☐ ☐

6. Is retraining conducted when changes in the workplace render previous training obsolete?
[29 CFR 1926.503(c)(1)]   ☐ ☐ ☐

7. Is retraining conducted when changes in the types of fall protection systems or equipment render previous training obsolete?
[29 CFR 1926.503(c)(2)]   ☐ ☐ ☐

8. Is retraining conducted if a person cannot use fall protection systems or equipment or has not retained the requisite understanding or skill?
[29 CFR 1926.503(c)(3)]   ☐ ☐ ☐

9.2 Personal Fall Protection Arrest Systems

Personal fall protection equipment that prevents a fall, such as a work restraint system, should always take priority over personal equipment which only limits the height or consequences of a fall (or both), such as a fall arrest system. MAKE SURE that

- Fall arrest system is suitable for the particular circumstances of the task—such as a restraint system should not be used for fall arrest.
- Selected system can be used for the particular task within its design limits, such as having an adequate clearance distance when using fall arrest equipment.
- Selected system meets the standard relevant to its intended use, OSHA for a full-body harness.

- Components are compatible, so the safe function of any one component is not adversely affected by, and does not interfere with, that of another; that is, DO NOT use an energy-absorbing lanyard to extend an inertia reel as this will affect the inertia reel's performance.

Maintenance guidelines

Basic care for fall protection safety equipment will prolong and endure the life of the equipment and contribute toward the performance of its vital safety function.

Proper storage and maintenance, after use, is as important as cleaning the equipment of dirt, corrosives, or contaminants.

- MAKE SURE the storage area is clean, dry, and free of exposure to fumes or corrosive elements.
- Keep equipment: clean and dry and properly stored.
- Dry equipment before storage, if it has become wet.

Note:
   Wipe off all dirt on nylon and polyester surfaces with a sponge dampened in plain water.
   Squeeze the sponge dry and dip the sponge in a mild solution of water and commercial soap or detergent.
   Work up a thick lather with a vigorous back-and-forth motion.
   Wipe the belt dry with a clean cloth. Hang freely to dry but away from excessive heat.
   Dry harnesses, belts, and other equipment thoroughly.
   DO NOT expose to heat, steam, or long periods of sunlight.

- Alter or repair equipment—only with the manufacturer approval.
- AVOID exposing fall protection arrest equipment and systems to damaging conditions, such as
  - Excessive heat that causes nylon to become brittle, resulting in a shriveled brownish appearance

Note: Fibers will break when flexed and should not be used above 180°F temperature.

  - Chemicals that produce a change in color, usually a brownish smear or smudge, with transverse cracks appearing when belt is bent over tight, causing a loss of elasticity in the belt
  - Ultraviolet rays, on webbing and rope lanyards stored in direct sunlight, resulting in a reduction of strength in some materials

- Molten metal or flame causing webbing and rope strands to fuse together

Note: Watch for hard, shiny spots or a hard and brittle feel; webbing will not support combustion, but nylon will.

- Paint and solvents penetrate and dry restricting movements of fibers.

Note: Drying agents and solvents in some paints will appear as chemical damage.
Source: The St. Paul, Tie or Die Fall Protection Program manual.

REGULATIONS	YES	NO	N/A
1. Are workers on scaffolds that are more than 10 ft (3.05 m) above a lower level protected from falling by one of the following measures:			
• Personal fall arrest system for workers on ladder jack scaffolds? [29 CFR 1926.451(g)(1)(i)]	☐	☐	☐
• Guardrail system installed within 9.5 in (24.13 cm) of and along at least one side of the walkway for workers on a walkway located within a scaffold? [29CFR 1926.451(g)(1)(v)]	☐	☐	☐
• Personal fall arrest system or guardrail system to protect workers doing overhand bricklaying from a supported scaffold from falling off open sides and ends of the scaffold (except at the side next to the wall being laid)? [29 CFR 1926.451(g)(1)(vi)]	☐	☐	☐
• Personal fall arrest system or guardrail system for workers on all other scaffolds? [29 CFR 1926.451(g)(1)(vii)]	☐	☐	☐
2. Does a competent person determine the feasibility and safety of providing fall protection for workers erecting or dismantling supported scaffolds? [29 CFR 1926.451(g)(2)]	☐	☐	☐
3. Are personal fall arrest systems, designed and installed by a qualified person, to maintain a safety factor of at least 2?	☐	☐	☐

4. Do personal fall arrest systems, and their use, comply with the provisions set forth in [29 CFR 1926.502(d)(1)], including
   - Are workers PROHIBITED from using body belts as part of a personal fall arrest system? ☐ ☐ ☐

Note: The use of a body belt in a positioning device system is acceptable and is regulated under [29 CFR 1926.502 (e)].

   - Are fall system connectors drop-forged, pressed or formed steel, or made of equivalent materials? [29 CFR 1926.502(d)(2)] ☐ ☐ ☐

5. Do workers erecting or dismantling supported scaffolds use fall protection when it is safe and feasible? [29 CFR 1926.451(g)(2)] ☐ ☐ ☐

LIFELINES AND LANYARDS   YES   NO   N/A

1. Are personal fall arrest systems used on scaffolds attached by a lanyard to a vertical lifeline, horizontal lifeline, or scaffold structural member? [29 CFR 1926.451(g)(3)] ☐ ☐ ☐

2. Are lifelines protected against being cut or abraded? [29 CFR 1926.502(d)(11)] ☐ ☐ ☐

3. Are horizontal lifelines designed, installed, and used (under the supervision of a qualified person) as part of a complete personal fall arrest system that maintains safety factor of at least 2? [29 CFR 1926.502(d)(8)] ☐ ☐ ☐

4. Are horizontal lifelines secured to two or more structural members of the scaffold? [29 CFR 1926.451(g)(3)(ii)] ☐ ☐ ☐

5. When vertical lifelines are used, is each person attached to a separate lifeline? [29 CFR 1926.502(d)(i)] ☐ ☐ ☐

6. Are vertical lifelines fastened to a fixed safe point of anchorage, independent of the scaffold, and protected from sharp edges and abrasion? [29 CFR 1926.451(g)(3)(i)] ☐ ☐ ☐

Note: Safe points of anchorage include structural members of buildings, but DO NOT include standpipes, vents, other piping systems, electrical conduit, outrigger beams, or counterweights.

7. Are workers PROHIBITED from attaching vertical lifelines and independent support lines to one another, to the same point of anchorage, and to the same point on the scaffold or personal fall arrest system?
   [29 CFR 1926.451(g)(3)(iv)] ☐ ☐ ☐

8. Do lanyards and vertical lifelines have minimum breaking strength of 5000 lb (2267.96 kg)?
   [29 CFR 1926.502(d)(9)] ☐ ☐ ☐

9. Can self-retracting lifelines and lanyards that automatically limit free-fall distance to 2 ft (60.96 cm) or less sustain tensile load of at least 3000 lb (1360.78 kg)—applied to the device with the lifeline or lanyard fully extended?
   [29 CFR 1926.502(d)(12)] ☐ ☐ ☐

10. Can the following equipment sustain a tensile load of at least 5000 lb (2267.96 kg), applied to the device with the lifeline or lanyard fully extended:
    - Self-retracting lifelines and lanyards that do not limit free-fall distance to 2 ft (60.96 cm) or less? ☐ ☐ ☐
    - Topstitch lanyards? ☐ ☐ ☐
    - Tearing and deforming lanyards? ☐ ☐ ☐
    [29 CFR 1926.502(d)(13)]

11. Are only ropes and straps (webbing), made of synthetic fibers, used in lanyards, lifelines, and strength components of body belts and body harnesses?
    [29 CFR 1926.502(d)(14)] ☐ ☐ ☐

12. Are anchorages used to attach personal fall arrest equipment separate from any anchorage used to support or suspend platforms? ☐ ☐ ☐

13. Can anchorages support at least 5000 lb (2267.96 kg) per person attached?
    [29 CFR 1926.502(d)(15)] ☐ ☐ ☐

Note: As an alternative, anchorages may be designed, installed, and used as part of a complete personal fall arrest system that maintains a safety factor of at least 2 and is under the supervision of a qualified person.

14. Are personal fall arrest systems, when used with a ☐ ☐ ☐
    body harness, able to limit the maximum arresting
    force on a person to 1800 lb (816.466 kg)?
    [29 CFR 1926.502(d)(16)(ii)]

15. Are personal fall arrest systems rigged such that, ☐ ☐ ☐
    when stopping a fall, a person can neither free-fall
    more than 6 ft (1.829 m), nor contact any lower
    level?
    [29 CFR 1926.502(d)(16)(iii)]

16. Do personal fall arrest systems bring a person to a ☐ ☐ ☐
    complete stop, when stopping a fall, and limit
    maximum deceleration distance to 3.5 ft (1.067 m)?
    [29 CFR 1926.502(d)(16)(iv)]

17. Can personal fall arrest systems withstand twice the ☐ ☐ ☐
    potential impact energy of a person, when stopping
    a free falling distance of 6 ft (1.829 m) or the free
    fall distance permitted by the system (whichever
    is less)?
    [29 CFR 1926.502(d)(16)(v)]

Note: The system is considered to be in compliance with the requirements outlined in questions 16 through 19 if

- Personal fall arrest system meets the criteria and protocols in Appendix C to subpart M, and
- System is used by an employee with a combined person and tool weight less than 310 lb (140 kg).

If the system is used by an employee having a combined tool and body weight of 310 lb (140 kg) or more, the employer must modify the criteria and protocols of the Appendix to provide proper protection for such heavier weights, or the system will not comply with the OSHA requirements.

18. Is the attachment point of the body harness located ☐ ☐ ☐
    in the center of the wearer's back near shoulder
    level or above the wearer's head?
    [29 CFR 1926.502(d)(17)]

19. Are body harnesses and components used only for ☐ ☐ ☐
    personal protection (as part of a personal fall arrest
    systems) and not to hoist materials?
    [29 CFR 1926.502(d)(18)]

	YES	NO	N/A

20. Are personal fall arrest systems and components that are subjected to impact loading immediately removed from service and not used again for protection until inspected by a competent person and determined to be undamaged and suitable for reuse?
[29 CFR 1926.502(d)(19)]

21. Are provisions made for prompt rescue in the event of a fall, or are employees able to rescue themselves?
[29 CFR 1926.502(d)(20)]

22. Are personal fall arrest systems inspected before each use for wear and damage, and are defective components removed from service?
[29 CFR 1926.502(d)(21)]

23. Are workers PROHIBITED from attaching personal fall arrest systems to guardrail systems or hoists?
[29 CFR 1926.502(d)(23)]

Note: OSHA regulations permit some exemptions.

24. Are personal fall arrest systems, used at hoist areas, rigged to allow the movement of the person only as far as the edge of the walking or working surface?
[29 CFR 1926.502(d)(24)]

ARREST SYSTEM CONNECTORS   YES   NO   N/A

1. Are connectors on personal fall arrest systems made of drop-forged, pressed, or formed steel, or of equivalent materials?
[29 CFR 1926.502(d)(1)]

2. Are connectors on personal fall arrest systems covered with a corrosion-resistant finish?
[29 CFR 1926.502(d)(3)]

3. Are all surfaces and edges smooth to prevent damage to interfacing parts of the system?
[29 CFR 1926.502(d)(2)]

4. Do D-rings and snap hooks on personal fall arrest systems have a minimum tensile strength of 5000 lb (2267.96 kg)?

5. Have D-rings and snap hooks on personal fall arrest systems been proof-tested to a minimum tensile load of 3600 lb (1632.93 kg) without cracking, breaking, or becoming deformed?
[29 CFR 1926.502(d)(4)]

6. Do locking snap hooks prevent disengagement of the snap hook if the connected member contacts the snap hook keeper?
[29 CFR 1926.502(d)(5)]

☐ ☐ ☐

7. Are locking snap hooks PROHIBITED from being used directly for the following types of connections (unless designed for that purpose):
   - Webbing, rope, or wire rope to each other?
   - D-ring, to which another snap hook or other connector is attached?
   - Horizontal lifeline?
   - Objects that are incompatible with the snap hook, such that an unintentional disengagement could occur?

   [29 CFR 1926.502(d)(6)]

☐ ☐ ☐
☐ ☐ ☐
☐ ☐ ☐
☐ ☐ ☐

WARNING LINE SYSTEMS YES NO N/A

1. Is a warning line erected around all sides of the roof work area?
[29 CFR 1926.502(f)(1)]

☐ ☐ ☐

2. When mechanical equipment is NOT used, is the warning line erected 6 ft (1.83 m) or more from the roof edge?
[29 CFR 1926.502(f)(1)(i)]

☐ ☐ ☐

3. When mechanical equipment IS used, is the warning line erected 6 ft (1.83 m) or more from the roof edge that is parallel to the direction of mechanical equipment operation, and 10 ft (3.05 m) or more from the roof edge that is perpendicular to the direction of mechanical equipment operation?
[29 CFR 1926.502(f)(1)(ii)]

☐ ☐ ☐

4. Are points of access, materials handling areas, storage areas, and hoisting areas connected to the work area by an access path formed by two warning lines?
[29 CFR 1926.502(f)(1)(iii)]

☐ ☐ ☐

5. When the path to a point of access is not in use, is a rope, wire, chain, or other barricade (equivalent in strength and height to the warning line) placed across the path, or is the path offset such that a person cannot walk directly into the work area?
[29 CFR 1926.502(f)(1)(iv)]

☐ ☐ ☐

Inspections

Note: Place the barricade at the point where the path intersects the warning line erected around the work area.

6. Do warning lines consist of supporting stanchions and ropes, wires, or chains?
   [29 CFR 1926.502(f)(2)] ☐ ☐ ☐

7. Is rope, wire, or chain flagged at 6-ft (1.83-m) intervals (or less) with high-visibility material?
   [29 CFR 1926.502(f)(2)(i)] ☐ ☐ ☐

8. Is the rope, wire, or chain rigged and supported so that its lowest point (including sag) is 34 in (86.36 cm) or more from the walking or working surface, and its highest point is 39 in (99.06 cm) or less from the walking or working surface?
   [29 CFR 1926.502(f)(2)(ii)] ☐ ☐ ☐

9. Can stanchions, after being erected (with the rope, wire, or chain attached), resist (without tipping over) a force that is
   - At least 16 lb (7.257 kg) applied horizontally against the stanchion? ☐ ☐ ☐
   - 30 in (76.2 cm) above the walking or working surface? ☐ ☐ ☐
   - Perpendicular to the warning line? ☐ ☐ ☐
   - In the direction of the floor, roof, or platform edge? ☐ ☐ ☐
     [29 CFR 1926.502(f)(2)(iii)]

10. Does the rope, wire, or chain have a minimum tensile strength of 500 lb (226.796 kg)?
    [29 CFR 1926.502(f)(2)(iv)] ☐ ☐ ☐

11. Can a stanchion, after the rope is attached to it, support (without breaking) the loads applied to the stanchions (as described in the preceding question 9)?
    [29 CFR 1926.502(f)(2)(iv)] ☐ ☐ ☐

12. Is the line attached at each stanchion so that pulling on one section of the line between stanchions will not result in slack being taken up in adjacent sections before the stanchion tips over?
    [29 CFR 1926.502(f)(2)(v)] ☐ ☐ ☐

13. Are employees PROHIBITED from entering the area between a roof edge and a warning line unless the worker is performing roofing work in that area?
    [29 CFR 1926.502(f)(3)] ☐ ☐ ☐

14. Is mechanical equipment on roofs used or stored only in areas where workers are protected by a warning line system, guardrail system, or personal fall arrest system?
[29 CFR 1926.502(f)(4)]   ☐ ☐ ☐

HARNESSES   YES NO N/A

Belts and Rings

1. Are there frayed edges, broken fibers, pulled stitches, cuts, or chemical damage?   ☐ ☐ ☐
2. Are D-rings and D-ring metal wear pads distorted, cracked, broken, and have rough or sharp edges?   ☐ ☐ ☐

Note: The D-ring bar should be at a 90-degree angle with the long axis of the belt and should pivot freely. PAY SPECIAL attention to attachments of buckles and D-rings.

3. Are there any unusual wear, frayed or cut fibers, or distortion of the buckles?   ☐ ☐ ☐
4. Are all rivets tight and unremovable with fingers?   ☐ ☐ ☐
5. Are body side rivet base and outside rivets flat against the material?   ☐ ☐ ☐

Note: Bent rivets will fail under stress.

6. Are there any frayed or broken webbing strands in the belt?   ☐ ☐ ☐

Note: Broken webbing strands generally appear as tufts on the webbing surface; any broken, cut, or burnt stitches will be readily seen.

Tongue Buckle

7. Are buckle tongues free of distortion in shape and motion?   ☐ ☐ ☐
8. Do they overlap the buckle frame and move freely back and forth in their socket?   ☐ ☐ ☐
9. Do rollers turn freely on the frame?   ☐ ☐ ☐
10. Are there any distortions or sharp edges?   ☐ ☐ ☐

Friction Buckle

11. Is there any distortion of the buckle? ☐ ☐ ☐

Note: Outer bar or center bars must be straight.

12. Is there any damage to corners and attachment points of the center bar? ☐ ☐ ☐

Lanyard

13. When the lanyard is slowly rotated, is there any distortion in its circumference? ☐ ☐ ☐
14. Is there any damage to spliced ends? ☐ ☐ ☐

HARDWARE	YES	NO	N/A

Snaps

1. Are there any hook and eye distortions, cracks, corrosion, or pitted surfaces? ☐ ☐ ☐
2. Does the keeper or latch seat into the nose without binding, distortion, or obstruction? ☐ ☐ ☐
3. Does the keeper spring exert sufficient force to firmly close the keeper? ☐ ☐ ☐

Note: **MAKE SURE** the keeper locks to prevent the keeper from opening when the keeper is closed.

Thimbles

4. Is the thimble (protective plastic sleeve) firmly seated in the eyes of splices? ☐ ☐ ☐
5. Does the splice have loose or cut strands? ☐ ☐ ☐
6. Are the edges of the thimble free of sharp edges, distortion, or cracks? ☐ ☐ ☐

LANYARDS	YES	NO	N/A

Steel Lanyard

1. When the lanyard is rotated: ☐ ☐ ☐
    - Are there cuts, frayed areas, or unusual wear patterns on the wire? ☐ ☐ ☐
    - Is a shock-absorbing device used when a steel lanyard is used for fall protection? ☐ ☐ ☐

Web Lanyard

2. When each side of a web lanyard is checked (bent over a piece of pipe),
   - Are there any cuts or breaks in the web? ☐ ☐ ☐
   - Is a shock-absorbing device used when a web lanyard is used for fall protection (because of a web lanyards' limited elasticity)? ☐ ☐ ☐

Rope Lanyard

3. When the lanyard is rotated:
   - Are there any fuzzy, worn, broken, or cut fibers? ☐ ☐ ☐

Note: Weakened areas from extreme loads will appear as a noticeable change in original diameter.

   - Is the rope diameter uniform throughout (following a short break-in period)? ☐ ☐ ☐
   - Is a shock-absorbing device used when a rope lanyard is used for fall protection? ☐ ☐ ☐

Webbing and Rope Lanyard Damage

4. Do webbing and rope lanyards have any damage caused by the following:
   - Ultraviolet rays—Exposure to direct sunlight reducing the strength of some materials? ☐ ☐ ☐
   - Heat—Exposure to heat above 180°F causing nylon to become brittle and shriveled with a brownish appearance and fibers breaking when flexed? ☐ ☐ ☐
   - Chemicals—Change in color, usually appearing as a brownish smear or smudge, with transverse cracks appearing when belt is bent over tight, causing a loss of elasticity in the belt? ☐ ☐ ☐
   - Paint and solvents—Penetration and drying that restrict movements of fibers? ☐ ☐ ☐

Note: Drying agents and solvents in some paints will appear as chemical damage.

   - Molten metal or flame—Fusing together of webbing and rope strands by molten metal or flame? ☐ ☐ ☐

Note: CHECK for hard, shiny spots or hard and brittle feel (webbing will not support combustion, but nylon will).

SHOCK-ABSORBING PACKS   YES   NO   N/A

1. Does the outer portion of the shock-absorbing pack have any burn holes or tears? ☐ ☐ ☐
2. Has the stitching, on areas where the pack is sewn to D-ring, belt, or lanyard, any loose strands, rips, and deterioration? ☐ ☐ ☐

SAFETY MONITORING SYSTEMS   YES   NO   N/A

1. Do safety monitoring systems, and their use, comply with the provisions of 1926.501(b)(10) and 1926.502(k)? ☐ ☐ ☐
2. Has the employer designated a competent person to monitor the safety of other employees? ☐ ☐ ☐
3. Does the employer ensure that the safety monitor complies with the following requirements:
   - Safety monitor MUST be competent to recognize fall hazards? ☐ ☐ ☐
   - Safety monitor warns the employee when it appears that the employee is unaware of a fall hazard or is acting in an unsafe manner? ☐ ☐ ☐
   - Safety monitor is standing on the same walking/working surface and within visual sighting distance of the employee being monitored? ☐ ☐ ☐
   - Safety monitor is close enough to communicate orally with the employee? ☐ ☐ ☐
   - Safety monitor DOES NOT have other responsibilities that could take the monitor's attention away from the monitoring function? ☐ ☐ ☐
   - Mechanical equipment MUST NOT be used or stored in areas where safety monitoring systems are being used to monitor employees engaged in roofing operations on low slope? ☐ ☐ ☐
4. Are employees, other than an employee engaged in roofing work (on low-sloped roofs) or an employee covered by a fall protection plan, PROHIBITED in an area where another employee or employees are being protected by a safety monitoring system? ☐ ☐ ☐

9.3 Safety Net Systems

Construction netting, both horizontal and vertical, is generally required on projects rising at least six stories or 75 ft (22.86 m), primarily to prevent tools, planks, pieces of concrete, and other debris from falling to the street, and threatening lives.

Horizontal netting, designed to protect workers in event of a fall, MUST be installed no more than two floors below where work is being done, generally protruding 10 ft (3.05 m) around the perimeter of construction sites, and must be raised as the building rises.

Vertical mesh is wrapped around floors that are framed but not yet enclosed by exterior walls. OSHA requires, where workers on a construction site are exposed to vertical drops of 6 ft (1.83 m) or more, that employers to provide fall protection in one of three ways before work begins:

- Placing guardrails around the hazard area
- Installing safety nets
- Providing personal fall arrest systems for each employee

Many times the nature and location of the work will dictate the form that fall protection takes. If the employer chooses to use a safety net system, it must comply with the following provisions:

- Safety nets must be installed as close as practicable under the surface on which employees are working, but in no case more than 30 ft (9.144 m) below.
- When nets are used on bridges, the potential fall area must be unobstructed.
- Safety nets must extend outward from the outermost projection of the work surface as follows.
- Safety nets must be installed with sufficient clearance to prevent contact with the surface or structures under them when subjected to an impact force equal to the drop test as described later.
- Safety nets and their installations must be capable of absorbing an impact force equal to the drop test as described below.
- Safety nets and safety net installations MUST be drop-tested at the jobsite:
  - After initial installation and before being used
  - Whenever relocated
  - After major repair
  - At 6-month intervals if left in one place

Notes: OSHA specifies a drop test consisting of a 400 lb (181.44 kg) bag of sand 28 to 32 in (71.82–81.28 cm) in diameter dropped into the net from the highest surface at which employees are exposed to fall hazards, but not from less than 42 in (1.07 m) above that level. When the employer can demonstrate that it is unreasonable to perform the drop-test as described earlier, the employer or a designated competent person shall certify that the net and net installation have sufficient clearance and impact absorption by preparing a certification record prior to the net being used as a fall protection system.

- CHECK that drop test certification includes
  - Identification of the net and net installation
  - Date when it was determined that the net and net installation were in compliance
  - Signature of the person making the determination and certification
- MAKE SURE the most recent certification record for each net and net installation MUST be available at the jobsite for inspection.

Maintenance guidelines

- PROVIDE care, maintenance, and storage of safety nets—in accordance with manufacturer's recommendations, at least once a week, giving particular attention to factors affecting net life:
  - Wear
  - Damage
  - Other deterioration
  - Any occurrence that could affect the integrity of the system

Note: Safety nets are capable of a minimum service life of 2 years under normal on-the-job exposure to weather, sunlight, and handling, excluding damage from misuse, mishandling, and exposure to chemicals and airborne contaminants.

- DO NOT use defective nets.
- REMOVE defective components from service.
- REMOVE, as soon as possible, objects and debris that fall into the safety net, such as scrap pieces, equipment, and tools—at least before the next work shift.
- VERIFY that maximum mesh size DOES NOT exceed 6 by 6 in (15.24 × 15.24 cm).
- MAKE SURE that all mesh crossings are secured to prevent enlargement of the mesh opening, which must be no longer than 6 in (15.24 cm), measured center-to-center.

- MAKE SURE each safety net, or section of net, has a border rope for webbing with a minimum breaking strength of 5000 lb (2267.96 kg).
- VERIFY that connections between safety net panels are as strong as integral net components and are NOT spaced more than 6 in (15.24 cm) apart.
- DO NOT use safety netting for storing materials.
- VERIFY that structural net hardware shall be drop-forged, pressed or formed steel, or material of equal or better quality.
- CHECK to see if surface is smooth and free of sharp edges.
- VERIFY that all hardware has a corrosion-resistant finish capable of withstanding a 50-hour salt spray test in accordance with ASTM B-1117.
- VERIFY that each structural net is permanently labeled with the following information:
  - Name of manufacturer
  - Identification of net material
  - Date of manufacturer
  - Date of prototype test
  - Name of testing agency
  - Serial number

SAFETY NET REGULATIONS YES NO N/A

1. Is a certification record available stating a competent person has inspected untested nets? ☐ ☐ ☐
2. Are safety nets extended beyond the edges of the walking/working surfaces 8 to 13 ft (2.44–3.96 m) depending on the potential fall distance? ☐ ☐ ☐
3. Are safety nets extended outward from the outermost projection of the work surface as follows:

Vertical distance from working level to horizontal plane of net	Minimum required horizontal distance of outer edge of net			
Up to 5 ft (1.524 m)	8 ft (2.428 m)	☐	☐	☐
5 ft (1.525 m) to 10 feet (3.048 m)	10 ft (3.048 m)	☐	☐	☐
More than 10 ft (3.048 m) [20 CFR 1926.502(c)(3)]	13 ft (3.962 m)	☐	☐	☐

4. Are safety nets installed as close as practicable under ☐ ☐ ☐
   the walking/working surface on which employees are
   working, but in no case more than 30 ft (9.1 m)
   below such level?
   [29 CFR 1926.502(c)(2)]

Note: When nets are used on bridges, the potential fall area from the walking/working surface to the net shall be unobstructed.

5. Are safety nets installed with sufficient clearance ☐ ☐ ☐
   under them to prevent contact with the surface or
   structures below when subjected to an impact force
   equal to the drop test specified in paragraph (c)(4)
   of this section?
   [29 CFR 1926.502(c)(4)]

6. Is the potential fall area from the walking/working ☐ ☐ ☐
   surfaces on bridges to nets unobstructed?

7. Are safety nets and their installations capable of ☐ ☐ ☐
   absorbing an impact force equal to that produced by
   the drop test specified in paragraph (c)(4)(i) of this
   section?
   [20 CFR 1926.502(c)(4)(i)]

Note: EXCEPT as provided in paragraph (2)(4)(ii) of this section.

8. Are safety nets and safety net installations
   drop-tested at the jobsite
   - After initial installation? ☐ ☐ ☐
   - Before being used as a fall protection system, ☐ ☐ ☐
     whenever relocated?
   - After major repair? ☐ ☐ ☐
   - At 6-month intervals if left in one place? ☐ ☐ ☐
   [29 CFR 1926.502(c)(4)(ii)]

Note: DROP TEST consists of a 400-lb (180-kg) bag of sand 30 plus or minus 2 in (76 + or − 5 cm) in diameter dropped into the net from the highest walking/working surface at which employees are exposed to fall hazards, but from not less than 42 in (1.1 m) above that level.

When the employer can demonstrate that it is unreasonable to perform the drop-test required by paragraph (c)(4)(i) of this section, the employer (or a designated competent person) shall certify that the net and net installation is in compliance with the provisions of paragraphs (c)(3) and (c)(4)(i) of this section by preparing a certification record prior to the net being used as a fall protection system.

The certification record MUST include an identification of the net and net installation for which the certification record is being prepared, the date when it was determined that the identified net and net installation were in compliance with paragraph (c)(3) of this section, and the signature of the person making the determination and certification.

The most recent certification record for each net and net installation MUST be available at the jobsite for inspection.
[29 CFR 1926.502(c)(5)]

9. Are workers PROHIBITED from using defective nets? ☐ ☐ ☐
[29 CFR 1926.502(c)(6)]

10. Are safety nets inspected at least once a week for ☐ ☐ ☐
wear, damage, and other deterioration with defective components removed immediately from service?
[29 CFR 1926.502(c)(6)]

11. Are safety nets also inspected after any occurrence ☐ ☐ ☐
which could affect the integrity of the safety net system?
[29 CFR 1926.502(c)(6)]

12. Do workers REMOVE from the safety net—as soon ☐ ☐ ☐
as possible—any materials, scrap pieces, equipment, and tools that have fallen into the net—and at least before the next work shift?
[29 CFR 1926.502(c)(7)]

13. Does maximum size of each safety net mesh opening ☐ ☐ ☐
NOT exceed 36 in^2 (232.26 cm^2) nor be longer than 6 in (15.24 cm) on any side, and the opening, measured center-to-center of mesh ropes or webbing, shall not be longer than 6 in (15.24 cm)?
[29 CFR 1926.502(c)(8)]

14. Are all mesh crossings secured to prevent enlargement ☐ ☐ ☐
of the mesh opening?
[29 CFR 1926.502(c)(8)]

15. Does each safety net (or section of it) shall have a ☐ ☐ ☐
border rope for webbing with a minimum breaking strength of 5000 lb (2267.96 kg)?
[29 CFR 1926.502(c)(9)

16. Are connections between safety net panels as strong as integral net components? ☐ ☐ ☐
17. Are connections spaced not more than 6 in (15.24 cm) apart? ☐ ☐ ☐
    [29 CFR 1926.502(d)]

9.4 Regulations

29 CFR 1910.66
    Subpart I   Personal Fall Arrest Systems
    Appendix C   Personal fall arrest system
    (Section I—Mandatory; Sections II and III—Non-mandatory)

29 CFR 1915.159
    Subpart I   Personal Fall Arrest systems (PFAS)

29 CFR 1926.501   Safety Standards for Fall Protection
    Subpart E   Personal Protective and Life-Saving Equipment
    .104   Safety belts, lifelines, and lanyards
    .105   Safety nets

29 CFR 1926.502
    Subpart M   Fall Protection System Criteria and Practices
    .502   Fall protection system criteria and practices
    .502(c)   Safety net systems
    .503   Training requirements

29 CFR 1926
    Subpart R   Steel Erection
    .759   Falling object protection
    .760   Fall protection

    Appendix B   Guardrail systems
Non-mandatory guidelines for complying with 1926.502(b)
    Appendix C   Personal fall arrest systems
Non-mandatory guidelines for complying with 1926.502(d)
    Appendix D   Positioning device systems
Non-mandatory guidelines for complying with 1926.502(e)

CSA Z259.10-M90   Fall Arresting Devices, Personnel Lowering Devices, and Lifelines

9.5 Standards (Consensus)

ANSI Z359.1-1992   Personal Fall Arrest Systems, Subsystems, and Components

Safety requirements for the performance, design, marking, qualification, instruction, training, inspection, use, maintenance, and removal from service of connectors, full-body harnesses, lanyards, energy absorbers, anchorage connectors, fall arresters, vertical lifelines, and self-retracting lanyards that comprise personal fall arrest systems for users within the capacity range of 130 lb (59 kg) to 310 lb (140 kg).

This standard addresses

- Only personal fall arrest systems (PFAS) incorporating full-body harnesses
- Equipment used in occupations requiring personal protection against falls from heights and applies to the manufacturers, distributors, purchasers, and users of such equipment.

This standard DOES NOT address

- Body belts, window cleaner belts, chest-waist harnesses, and chest harnesses—even when referred to as body supports
- Construction industry and sports-related activities

All requirements of this standard MUST be met—before any equipment shall bear the marking Z539.1 or be represented in anyway as being in conformance with this standard.

Variance from the requirements of this standard is permissible in isolated instances of practical difficulties when applying it all the user level—BUT ONLY when it is clearly evident that an equivalent degree of protection is thereby secured.

ANSI/ASSE Z350   Standards for Fall Protection
Z359.1-2007	Safety Requirements for Personal Fall Arrest Systems, Subsystems, and Components
Z359.2-2007	Minimum Requirements for a Comprehensive Managed Fall Protection Program
Z359.3-2007	Safety Requirements for Positioning and Travel Restraint Systems
Z359.4-2007	Safety Requirements for Assisted-Rescue and Self-Rescue Systems, Subsystems, and Components
Z359.6-2009	Specifications and Design Requirements for Active Fall Protection Systems
Z359.12-2009	Connecting Components for Personal Fall Arrest System
Z359.13-2009	Personal Energy Absorbers and Energy-Absorbing Lanyards

ANSI/ASSE A10 Safety Requirements for Construction and Demolition Operations
- A10.11 Safety Nets
- A10.14 Safety Belt Harnesses, Lanyards, and Lifelines
- A10.21 Proper Cleaning and Disposal of Contaminated Work Clothing
- A10.32 Fall Protection Systems for Construction
- A10.34 Public Protection
- A10.37 Debris Nets

Chapter 10

Ladders, Stairways, and Ramps

Before working with a ladder for the first time, read the manufacturer's instructions.

> ALWAYS thoroughly inspect ladders, each time they are used (ANSI A14.1-2007).
>
> Check that rungs are firm and unbroken; braces are fastened securely; and ropes, pulleys, and other moving parts are in good working order.
>
> Remove damaged ladders from service.
>
> Repair damaged ladders before placing back in service.

Note: If repairs are not feasible, the defective ladder should be taken out of service and a new ladder be used in its place.

Fill out and attach ladder inspection tags, to the ladders, to ensure that the ladders have been inspected.

Maintenance guidelines

Use a good checklist when performing preventive maintenance to make sure all the work items are performed, and make notes of any issues that should be checked into further.

The three important aspects of ladder maintenance are regular cleaning, inspection, and lubrication regardless of whether timber, aluminum, or fiberglass. Follow all scaffolding manufacturer's recommendations—as well as all federal, state, and local regulations, codes, and ordinances pertaining to scaffolds and scaffolding systems.

- Check timber ladders to ensure that
  - They are clean, free of splinters, spilt paint, and other opaque deposits.
  - Their feet are securely fitted and in good condition—not excessively worn.
  - They are free from rot, decay, or mechanical damage such as warped stiles, excessive cracks, splintering, and wear and tear at the head and foot of the stiles.
  - There is NO looseness of rungs, excessive wear, or decay where the rung enters the stile.
  - Ropes are free of fraying, wear, rot, or other damage.
  - Steps or rungs are firmly fitted.
  - Steps or rungs provide good slip resistance.
  - All components are securely fitted and free of damage or corrosion.
  - There is NO longitudinal play.
  - All moving parts operate freely without bending or undue play.
- Check stepladders to ensure that
  - Ladder is in good condition and free from slippery substances.
  - Stays, chains, or cords are in good condition.
  - Hinges are firmly fitted.
  - Spreaders operate effectively.
  - Ladder stands firm; does not wobble.
- DO NOT paint wooden ladders with solid color paints.

Note: Paint may mask cracks in the wood and make them hard to see.

- Use only clear wood preservative to protect bare wood.
- Remove from service, immediately, any damaged ladders, and either repair or destroy them.
- Tag or mark ladders found to be defective or to have unsafe conditions, so that they will not be used.
- DO NOT risk using a faulty ladder until replacement parts have been fitted or loose joints or fittings have been secured.
- Check extension ladders to ensure that
  - Clutches, stops, guide irons, and pulleys are in good condition and operate correctly.
  - Lubrication of all working parts, pivoting or rotating surfaces, including pulleys, is regular and sufficient, with light machine oil.
- Check cords, chains, and ropes and replace if defective.

- Replace missing or defective pads or sleeves.
- Check metal ladders for twisting, distortion, oxidation, corrosion, and excessive wear, especially on treads.
- Replace properly any broken or loose rungs, defective tie rods and broken rivets, loose hinges, or other defective metal fittings.
- Check glass-reinforced plastic ladders for mechanical damage.
- Check ALL ladders to ensure that the following information is permanently marked on each ladder in a prominent position, in the largest letting practicable:
  - Name and city of the manufacturer

Warning: Ladder is intended to carry a maximum load of ... [load rating in lb (kg)].

- Working length of the ladder (closed and maximum working lengths for extension-type ladders)
- To be used in the fully open position only (on double-sided stepladders)
- DO NOT USE where electrical hazards exist (on wire-bound ladders)

Warning: DO NOT stand any higher than the second top tread on a stepladder that is NOT a platform stepladder.

- Take ladders out of service, if contaminated with oil, grease, mud, or other slippery substances, and clean them before returning into service.
- ALWAYS store ladders in a covered, ventilated area, protected from the weather and away from too much dampness or heat.
- Store ladders horizontally on a rack or wall brackets.

Note: Racks should have sufficient supporting points to prevent excessive sagging. Materials should not be placed on stored equipment.

- Secure the top of a ladder, if stored vertically, with a bracket to prevent it from falling.
- NEVER hang a ladder from a rung and don't store a ladder in any place where a child might be tempted to climb it.
- Store timber ladders off the ground to avoid contact with dampness, providing good ventilation.
- DO NOT store timber ladders near radiators, stoves, steam pipes, or in areas subject to excessive heat or dampness.

- Hang aluminum ladders horizontally from a stile or on a stile—but beware—long and heavy ladders can sag in the middle if not supported sufficiently.
- *Note:* This sagging cannot be repaired and renders the ladder unusable!
- DO NOT store a ladder outdoors for good maintenance as well as security; it could be stolen or used in a break-in.

Source: Lansford Access Ltd, Gloucester, UK. (www.igloucestershire.co.uk)

Safety guidelines

- DO NOT use a ladder if sleepy or ill, taking medication, or if bad weather conditions exist.
- DO NOT use ladders in doorways or other high-traffic areas.

Note: If a ladder must be used near a door, make sure the door is locked. If the door has to be open or the ladder is in a raised position, ask a coworker to stay with the ladder to make sure an accident does not occur.

Use fiberglass or wood ladders, rather than metal, near power lines or other sources of electricity to avoid electric shock hazards.

- SET the feet of a ladder level and position solidly on the ground.
- USE boards under the legs for support if the ground is soft or uneven.
- TEST the ladder to verify it is secure.
- MAKE SURE for stability that both sides of the ladder are set against the wall or other support.
- SPREAD the legs on a stepladder fully and locked into position.
- PLACE ladders at a 75-degree angle, as REQUIRED by regulations.
- MAKE SURE hands, shoes, and ladder rungs are dry.
- USE a second person to hold the bottom of the ladder and prevent others from disturbing ladder.
- KEEP a three-point grip on the ladder at all times (two hands and one foot or one hand and two feet).
- AVOID distractions that make you turn away from the front of the ladder.

- CLIMB SLOWLY with weight centered between side rails.
- DO NOT lean back.
- NEVER stand on the top two rungs of a stepladder or top four rungs of an extension ladder.

SAFETY INSPECTION	YES	NO	N/A
1. Are employees made aware that metal ladders are not to be used where they may contact electric conductors or equipment?	☐	☐	☐
2. Are employees aware that metal ladders cannot be used when working on or near electric equipment, such as changing light bulbs or fluorescent tubes?	☐	☐	☐
3. Are ladders with broken or missing rungs or split side rails tagged and taken out of service?	☐	☐	☐
4. Are metal ladders inspected for damage or signs of corrosion?	☐	☐	☐
5. Are portable wood ladders and metal ladders adequate for their purpose, in good condition, and provided with secure footing?	☐	☐	☐
6. Are areas around the top and bottom of the ladder kept clear?	☐	☐	☐
7. Are portable ladders used at such a pitch that the horizontal distance from the top bearing to the foot of the ladder is about one-fourth of the working length of the ladder?	☐	☐	☐
8. Are ladders prohibited from being used in a horizontal position as platforms, runways, or scaffolds?	☐	☐	☐
9. Are portable ladders tied, blocked, or otherwise secured against movement?	☐	☐	☐
10. Are the rungs of ladders uniformly spaced, if simultaneous two-way traffic is expected?	☐	☐	☐
11. Are the side rails of the ladder extending at least 36 in (91.44 cm) above the landing?	☐	☐	☐
12. Did a competent person provide ladder safety training to all employees?	☐	☐	☐

13. Are stairways or ladders provided at all breaks in elevation 19 in (48.26 cm) or higher to provide safe access for employees? ☐ ☐ ☐

14. Are stairways in good condition and are stair rails provided for stairways having four or more risers? ☐ ☐ ☐

Source: Maryland Occupational Safety & Health, Construction Checklist.

LADDER MAINTENANCE	YES	NO	N/A
1. Are all ladders maintained in good condition, joints between steps and side rails tight, all hardware and fittings securely attached, and moveable parts operating freely without binding or undue play?	☐	☐	☐
2. Are nonslip safety feet provided on all metal or rung ladders (except stepladders)?	☐	☐	☐
3. Are ladder rungs and steps free of grease and oil?	☐	☐	☐
4. Are employees PROHIBITED from placing a ladder in front of doors opening toward the ladder unless the door is blocked open, locked, or guarded?	☐	☐	☐
5. Are employees PROHIBITED from placing ladders on boxes, barrels, or other unstable bases to obtain additional height?	☐	☐	☐
6. Are employees instructed to face the ladder when ascending or descending?	☐	☐	☐
7. Are employees PROHIBITED from using ladders that are broken; have missing steps, rungs, or cleats; broken side rails; or other faulty equipment?	☐	☐	☐
8. Are employees instructed NOT to use the top step of ordinary stepladders as a step?	☐	☐	☐
9. Do portable rung ladders always extend at least 3 ft (91.44 cm) above the elevated surface, when used to gain access to elevated platforms, roofs, etc?	☐	☐	☐
10. Are employees required, when using portable rung or cleat-type ladders, to place the ladder so that slipping will not occur, or it is lashed or otherwise held in place?	☐	☐	☐

11. Are employees required to secure the base of a portable rung or cleat-type ladder to prevent slipping, or otherwise lash or hold it in place? ☐ ☐ ☐

12. Are portable metal ladders legibly marked with signs reading—CAUTION: DO NOT USE AROUND ELECTRICAL EQUIPMENT—or equivalent wording? ☐ ☐ ☐

13. Are employees prohibited from using ladders as guys, braces, skids, gin poles, or for other than their intended purposes? ☐ ☐ ☐

14. Are employees instructed to only adjust extension ladders while standing at a base (not while standing on the ladder or from a position above the ladder)? ☐ ☐ ☐

15. Are metal ladders inspected for damage? ☐ ☐ ☐

16. Are the rungs of ladders uniformly spaced at 12 in (30.48 cm) center to center? ☐ ☐ ☐

OSHA regulations

Occupational Safety and Health Administration (OSHA) regulations, under the general industry standard 29 CFR 1910.27, apply to fixed ladders between the pitch range 60 and 90 degrees of the horizontal. Fixed ladders in the range of 75 to 90 degrees of the horizontal, however, are preferred.

The regulations apply to wooden and metal ladders, including stepladders, but they DO NOT address fixed ladders with cages, baskets, wells, hatch covers, landing platforms, or ladders exceeding 20 ft in height (see 29 CFR 1910.27).

For fixed ladders associated with construction sites (see 29 CFR 1926.1051 and 1926.1052).

The regulations cited apply only to private employers and their employees—unless adopted by a state agency and applied to other groups such as public employees

10.1 General Industry Ladders

	YES	NO	N/A

1. Is the distance between rungs 12 in (30.48 cm) or less and uniform throughout the length of the ladder?
   [29 CFR 1910.27(b)(1)(ii)] ☐ ☐ ☐

2. Is the minimum clear length of rungs or cleats at least 16 in (40.68 cm)?
   [29 CFR 1910.27(b)(1)(iii)]   ☐ ☐ ☐

3. Are the rungs, cleats, and steps free of splinters, sharp edges, burrs, or other projections that are hazards?
   [29 CFR 1910.27(b)(1)(iv)]   ☐ ☐ ☐

4. Are the rungs of ladders designed so that the foot cannot slip off the end?
   [29 CFR 1910.27(b)(1)(v)]   ☐ ☐ ☐

5. Are the rungs of metal ladders at least three-fourths of 1 in (25.4 mm) in diameter?
   [29 CFR 1910.27(b)(1)(i)]   ☐ ☐ ☐

Note: If the ladder is embedded in concrete and serves as an access to pits or other atmospheres that cause corrosion and rusting, the rungs must be at least 1 in (25.4 cm) in diameter or the rungs MUST be painted or treated to resist corrosion and rusting [29 CFR 1910.27(b)(7)(i)].

6. Are the rungs of wood ladders at least 11.25 in (28.575 cm) in diameter?
   [29 CFR 1910.27(b)(1)(i)]   ☐ ☐ ☐

7. Do side rails that might be used as a climbing aid provide adequate gripping surface without sharp edges, splinters, or burrs?
   [29 CFR 1910.27(b)(2)]   ☐ ☐ ☐

8. Are metal ladders painted or treated to protect them from corrosion and rusting when location demands?
   [29 CFR 1910.27(b)(7)(i)]   ☐ ☐ ☐

9. Are wood ladders treated with a nonirritating preservative when used under conditions that may cause decay?
   [29 CFR 1910.27(b)(7)(ii)]   ☐ ☐ ☐

10. Is the perpendicular distance from the center line of the rungs to the nearest permanent object on the climbing side of the ladder 36 in (91.44 cm) for a pitch of 76 degrees and 30 in (76.2 cm) for a pitch of 90 degrees?
    [29 CFR 1910.27(c)(1)]   ☐ ☐ ☐

Note: The intent of this provision is to prevent the climber from bumping his or her head or shoulder on an object while climbing or descending the ladder.

11. Is the distance from the center line of the rung, ☐ ☐ ☐
    cleat, or step to the nearest permanent object
    behind the ladder at least 7 in (17.18 cm)?
    [29 CFR 1910.27(c)(4)]

Note: The purpose of this provision is to give adequate clearance so that the climber's foot does not strike an object between the wall and the ladder and cause a fall.

12. Is the distance from the center line of the grab bar ☐ ☐ ☐
    to the nearest permanent object behind the grab
    bar at least 4 in (10.16 cm)?
    [29 CFR 1910.27(c)(5)]

Note: The purpose of this provision is to ensure that the climber can grab the bar quickly and easily.

13. Is the step-across distance from the nearest edge ☐ ☐ ☐
    of the ladder to the nearest edge of the equipment
    or structure not more than 12 in (30.48 cm) and
    not less than 2.5 in (6.36 cm)?
    [29 CFR 1910.27(c)(6)]

Note: If the step-across distance is greater than 12 in (30.48 cm), a landing platform of at least 24 in (60.96 cm) wide and 30 in (76.2 cm) long must be provided with standard railings and toeboards. Consult the OSHA regulations 29 CFR 1910.27(d)(2)(ii) for requirements.

14. Are side rails of ladders extended at least 3.5 ft ☐ ☐ ☐
    (1.0668 m) above the landing?
    [29 CFR 1910.27(d)(3)]

Note: The purpose of this standard is to give the climber something to hold onto while getting off and onto the ladder.

15. Are all ladders inspected regularly and maintained ☐ ☐ ☐
    in a safe condition?
    [29 CFR 1910.27(f)]

10.2 Fixed Ladders

Regulations

This checklist covers regulations issued by the U.S. Department of Labor, Occupational Safety and Health Administration (OSHA) under the general industry standard 29 CFR 1910.27.

It applies to fixed ladders in the pitch range of 60 to 90 degrees with the horizontal.

- Preferred pitch—Fixed ladders are preferred to be in the range of 75 to 90 degrees with the horizontal
- Substandard pitch—Fixed ladders are considered as substandard if they are installed within the substandard pitch range of 60 to 75 degrees with the horizontal. (Substandard fixed ladders are permitted only where it is found necessary to meet conditions of installation. This substandard pitch range shall be considered as a critical range to be avoided, if possible.)
- Pitch greater than 90 degrees—Ladders are PROHIBITED from being put in use having a pitch in excess of 90 degrees with the horizontal.

The regulations cited apply only to private employers and their employees, unless adopted by a state agency and applied to other groups such as public employees.

This checklist does not address fixed ladders with cages, baskets, wells, hatch covers, landing platforms, or ladders exceeding 20 ft in height (see 29 CFR 1910.27).

It also does not address fixed ladders associated with construction sites. See section 10.5—Construction Site Ladders covering 29 CFR 1926.1051 and 1926.1053 regulations.

	YES	NO	N/A
1. Is the distance between rungs 12 in (30.48 cm) or less and uniform throughout the length of the ladder? [29 CFR 1910.27(b)(1)(ii)]	☐	☐	☐
2. Is the minimum clear length of rungs or cleats at least 16 in (40.64 cm)? [29 CFR 1910.27(b)(1)(iii)]	☐	☐	☐
3. Are the rungs, cleats, and steps free of splinters, sharp edges, burrs, or other projections that are hazards? [29 CFR 1910.27(b)(1)(iv)]	☐	☐	☐
4. Are the rungs of ladders designed so that the foot cannot slip off the end? [29 CFR 1910.27(b)(1)(v)]	☐	☐	☐
5. Are the rungs of metal ladders at least 0.75 in (19.05 mm) in diameter? [29 CFR 1910.27(b)(1)(i)]	☐	☐	☐

Note: If the ladder is embedded in concrete and serves as an access to pits or other atmospheres that cause corrosion and rusting, the rungs must be at least 1 in. in diameter or be painted or treated to resist corrosion and rusting.

6. Are the rungs of wood ladders at least 1.125 in (28.58 mm) diameter? ☐ ☐ ☐
[29 CFR 1910.27(b)(1)(i)]

7. Do side rails that might be used as a climbing aid provide adequate gripping surface without sharp edges, splinters, or burrs? ☐ ☐ ☐
[29 CFR 1910.27(b)(2)]

8. Are metal ladders painted or treated to protect them from corrosion and rusting when location demands? ☐ ☐ ☐
[29 CFR 1910.27(b)(7)(i)]

9. Are wood ladders treated with a nonirritating preservative when used under conditions that may cause decay? ☐ ☐ ☐
[29 CFR 1910.27(b)(7)(ii)]

10. Is the perpendicular distance from the center line of the rungs to the nearest permanent object on the climbing side of the ladder 36 in (91.44 cm) for a pitch of 76 degrees and 30 in (76.2 cm) for a pitch of 90 degrees.? ☐ ☐ ☐
[29 CFR 1910.27(c)(1)]

Note: The intent of this provision is to prevent the climber from bumping his or her head or shoulder on an object while climbing or descending the ladder.

11. Is the distance from the center line of the rung, cleat, or step to the nearest permanent object behind the ladder at least 7 in (17.78 cm)? ☐ ☐ ☐
[29 CFR 1910.27(c)(4)]

Note: The purpose of this provision is to give adequate clearance so that the climber's foot does not strike an object between the wall and the ladder and cause a fall.

12. Is the distance from the center line of the grab bar to the nearest permanent object behind the grab bar at least 4 in (10.16 cm)? ☐ ☐ ☐
[29 CFR 1910.27(c)(5)]

Note: The purpose of this provision is to ensure that the climber can grab the bar quickly and easily.

13. Is the step-across distance from the nearest edge ☐ ☐ ☐
    of the ladder to the nearest edge of the equipment/
    structure not more than 12 in (30.48 cm) and not
    less than 2.5 in (6.35 cm)?
    [29 CFR 1910.27(c)(6)]

Note: If the step-across distance is greater than 12 in (30.48 cm), a landing platform of at least 24 in (60.96 cm) wide and 30 in (76.2 cm) long must be provided with standard railings and toe boards. Consult the OSHA regulations 29 CFR 1910.27(d)(2)(ii) for requirements.

14. Are the side rails of ladders extended at least 3.5 ft ☐ ☐ ☐
    (1.07 m) above the landing?
    [29 CFR 1910.27(d)(3)]

Note: The purpose of this standard is to give the climber something to hold onto while getting off and onto the ladder.

15. Are all ladders inspected regularly and maintained ☐ ☐ ☐
    in a safe condition?
    [29 CFR 1910.27(f)]

10.3 Portable Ladders

Ladder accidents usually are caused by improper selection, care, or use, not by manufacturing defects. Some of the more common hazards involving ladders, such as instability, electric shock, and falls, can be predicted and prevented. Prevention requires proper planning, correct ladder selection, good work procedures, and adequate ladder maintenance.

Maintenance guidelines

Use a good checklist when performing preventive maintenance to make sure all the work items are performed, and make notes of any issues that should be checked into further.

- PROTECT wood ladders with a clear sealer varnish, shellac, linseed oil, or wood preservative.
- DO NOT paint wood ladders because the paint could hide defects.
- CHECK carefully for cracks, rot, splinters, broken rungs, loose joints and bolts, and hardware in poor condition.

- INSPECT aluminum, or steel ladders rough burrs and sharp edges before the use. Inspect closely for loose joints and bolts, faulty welds, and cracks.
- MAKE SURE the hooks and locks on extension ladders are in good condition. Replace worn or frayed ropes on extension ladders at once.
- MAINTAIN fiberglass ladders with a surface coat of lacquer. If it is scratched beyond normal wear, it should be lightly sanded before applying a coat of lacquer.
- ALWAYS inspect the ladder prior to using it. If the ladder is damaged, it must be removed from service and tagged until repaired or discarded.
- MAKE SURE ladders are free of any slippery material on the rungs, steps, or feet.
- CHECK for a side rail damage, which may cause one side of a ladder to give way.
- PROVIDE nonskid feet or spurs to prevent a ladder from slipping on a hard, smooth surface.
- MAKE SURE all locks on an extension ladder are properly engaged, holding to prevent overloading a rail.
- Replace the top rung of portable ladders with chain or rope to reduce rocking, when working on cylindrical objects like poles and columns.

Sources:
National Safety Council Publications
Job Made Ladders, Data Sheet No. 1-568-76, 1976
Accident Prevention Manual for Industrial Operations, 9th Edition, 1988
National Institute for Occupational Safety and Health

Portable wood ladders

This checklist covers regulations issued by the U.S. Department of Labor, Occupational Safety and Health Administration (OSHA) under the general industry standards 29 CFR 1910.25 (portable wooden ladders).

It applies to wooden and metal ladders, including stepladders.

It DOES NOT apply to stockroom stepladders, aisle-way stepladders, shelf ladders, and library ladders. The regulations cited apply only to private employers and their employees, unless adopted by a state agency and applied to other groups such as public employees.

Inspections

Use the checklist entitled Portable Ladders for Construction for construction site situations.

	YES	NO	N/A
1. Are all ladders maintained in good condition, joints between steps and side rails tight, all hardware and fittings securely attached, and moveable parts operating freely without binding or undue play?	☐	☐	☐
2. Are all wooden ladder parts sound and free of sharp edges and splinters? [29 CFR 1910.25(b)(1)(i)]	☐	☐	☐
3. Are they, on visual inspection, free from shake, wane, compression failure, decay, or other irregularities? [29 CFR 1910.25(b)(1)(i)]	☐	☐	☐
4. Are all portable wooden stepladders 20 ft (6.1 m) or less in length? [29 CFR 1910.25(c)(2)]	☐	☐	☐
5. Is the portable stepladder of uniform step spacing and less than 12 in (30.48 cm) apart? [29 CFR 1910.25(c)(2)(i)(b)]	☐	☐	☐
6. Is the inside width between side rails of each portable stepladder at least 11.5 in? [29 CFR 1910.25(c)(2)(i)(c)]	☐	☐	☐
7. Is the metal spreader or locking device of portable stepladders of sufficient size and strength to securely hold the front and back sections in the open position? [29 CFR 1910.25(c)(2)(i)(f)]	☐	☐	☐
8. Are all single wooden ladders 30 ft (9.15 m) or less in length? [29 CFR 1910.25(c)(3)(ii)(a)]	☐	☐	☐
9. Are all two-section wooden extension ladders 60 ft (12.29 cm) or less in length? [29 CFR 1910.25(c)(3)(iii)(a)]	☐	☐	☐

10. Are all wooden ladders in good condition with the joint between step and side rails tight?

	YES	NO	N/A
▪ Are all hardware and fittings securely attached?	☐	☐	☐
▪ Are the movable parts operating freely without binding or undue play? [29 CFR 1910.25(d)(1)(i)]	☐	☐	☐

11. Are the metal bearings of locks, wheels, pulleys, etc. frequently lubricated?
    [29 CFR 1910.25(d)(1)(ii)]   ☐ ☐ ☐

12. Is frayed or badly worn rope replaced?
    [29 CFR 1910.25(d)(1)(iii)]   ☐ ☐ ☐

13. Are the safety feet or other auxiliary equipment kept in good condition?
    [29 CFR 1910.25(d)(1)(iv)]   ☐ ☐ ☐

14. Are wooden ladders inspected frequently? Are those with defects withdrawn from service for repair or destruction and tagged or marked as DANGEROUS, DO NOT USE?
    [29 CFR 1910.25(d)(1)(x) and (d)(2)(viii)]   ☐ ☐ ☐

Note: Wooden ladders with missing steps, rungs, or cleats; broken side rails; or other faulty equipment MUST NOT be used. Discarded ladders should be cut down the center of the rungs.

15. Are rungs kept free of grease and oil?
    [29 CFR 1910.25(d)(1)(xi)]   ☐ ☐ ☐

16. Are wooden ladders used and placed so that the horizontal distance from the top support to the foot of the ladder is one-quarter of the working length of the ladder (the length along the ladder between the foot and the top support)?
    [29 CFR 1910.25(d)(2)(i)]   ☐ ☐ ☐

17. Is the ladder placed to prevent slipping, or lashed, or held in position?
    [29 CFR 1910.25(d)(2)(i)]   ☐ ☐ ☐

18. Is the use of wooden ladders in the horizontal position prohibited?
    [29 CFR 1910.25(d)(2)(i)]   ☐ ☐ ☐

Note: Ladders must never be used as platforms, runways, or scaffolds.

19. Is only one person allowed on the ladder at one time?
    [29 CFR 1910.25(d)(2)(ii)]   ☐ ☐ ☐

20. Are ladders placed away from the front of doors that open toward the ladder unless the door is blocked, locked, or guarded?
    [29 CFR 1910.25(d)(2)(iv)]   ☐ ☐ ☐

21. Are ladders always placed on stable bases? ☐ ☐ ☐
    - Are employees PROHIBITED from placing ladders on boxes, barrels, or other unstable bases to obtain additional height?
      [29 CFR 1910.25(d)(2)(v)] ☐ ☐ ☐
22. Is the splicing of short ladders together prohibited?
    [29 CFR 1910.25(d)(2)(ix)] ☐ ☐ ☐
23. Is the use of the tops of stepladders as steps prohibited?
    [29 CFR 1910.25(d)(2)(xii)] ☐ ☐ ☐
24. When in use, do all 36-ft (10.97-m) or less, two-section extension wooden ladders have a minimum overlap of 3 ft (91.44 cm) between the two sections?
    [29 CFR 1910.25(d)(2)(xiii)] ☐ ☐ ☐
25. When in use, do all 36-ft (10.97-m) to 48-ft (14.63-m) two-section extension wooden ladders have a minimum overlap of 4 ft (1.22 m) between the two sections?
    [29 CFR 1910.25(d)(2)(xiii)] ☐ ☐ ☐
26. When in use, do all 48-ft (14.63-m) to 60-ft (18.29-m) two-section extension wooden ladders have a minimum overlap of 5 ft (1.52 m) between the two sections?
    [29 CFR 1910.25(d)(2)(xiii)] ☐ ☐ ☐
27. If ladders are used to gain access to a roof, are they extended at least 3 ft (91.44 cm) above the point of support?
    [29 CFR 1910.25(d)(2)(xv)] ☐ ☐ ☐
28. Are all portable rung ladders equipped with nonslip bases where hazard of slipping exists?
    [29 CFR 1910.25(d)(2)(xix)] ☐ ☐ ☐

Note: Nonslip bases are not intended as a substitute for care in safely placing, lashing, or holding a ladder that is being used.

Portable metal ladders

This checklist covers regulations issued by the U.S. Department of Labor, Occupational Safety and Health Administration (OSHA) under the general industry standards 29 CFR 1910.26 (portable metal ladders).

It applies to metal ladders, including stepladders.

It DOES NOT apply to stockroom stepladders, aisle-way stepladders, shelf ladders, and library ladders. The regulations cited apply only to private employers and their employees, unless adopted by a state agency and applied to other groups such as public employees.

Use the checklist entitled Portable Ladders for Construction for construction site situations

	YES	NO	N/A
1. Are metal ladders maintained in good usable condition at all times? [29 CFR 1910.26(c)(2)(iv)]	☐	☐	☐
2. Are nonslip safety feet provided on each metal or rung ladder?	☐	☐	☐
3. Are the rungs and steps of portable metal ladders corrugated, knurled, dimpled, coated with skid-resistant material, or otherwise treated to minimize the possibility of slipping? [29 CFR 1910.26(a)(1)(v)]	☐	☐	☐
4. Are all portable metal single ladders 30 ft (9.144 m) or less in length? [29 CFR 1910.26(a)(2)(ii)]	☐	☐	☐
5. Are all portable metal two-section ladders 48 ft (14.63 m) or less in length? [29 CFR 1910.26(a)(2)(ii)]	☐	☐	☐
6. If a portable metal ladder tips over, is it inspected immediately for damage? [29 CFR 1910.26(c)(2)(vi)(a)]	☐	☐	☐

Note: The inspection must include looking for dents, bends, or excessively dented rungs and checking all rungs to side rail connections, hardware connections, and rivets for shears.

	YES	NO	N/A
7. Are metal ladders cleaned immediately if exposed to oil and grease? [29 CFR 1910.26(c)(2)(vi)(d)]	☐	☐	☐
8. Are metal ladders with defects marked and taken out of service until repaired by either the maintenance department or the manufacturer? [29 CFR 1910.26(c)(2)(vii)]	☐	☐	☐
9. Are metal ladders placed at the proper angle? [29 CFR 1910.26(c)(3)(i)]	☐	☐	☐

Note: That is, the base distance from the vertical wall to the ladder is one-fourth the working length of the ladder or height at which the ladder touches the wall.

10. Are workers PROHIBITED from using a metal ladder as a brace, skid, guy or gin pole, gangway, or for other uses than that which the ladder was intended?
[29 CFR 1910.26(c)(3)(vii)] ☐ ☐ ☐

11. Has inspection been conducted to determine if metal ladders might contact energized conductors?
[29 CFR 1910.26(c)(3)(viii)] ☐ ☐ ☐

Note: The use of metal ladders should be prohibited wherever they might make contact with energized electric conductors.

Source: NIOSH—National Institute for Occupational Safety and Health, Cincinnati, OH.

10.4 Mobile Ladder Stands

This checklist covers regulations issued by the U.S. Department of Labor, Occupational Safety and Health Administration (OSHA) under the general industry standard 29 CFR 1910.29.

It applies to mobile ladder stands that may be used in labs and shops to reach lights or other overhead storage areas.

The regulations cited apply only to private employers and their employees, unless adopted by a state agency and applied to other groups such as public employees.

	YES	NO	N/A

1. Are all exposed surfaces of mobile ladder stands free from sharp edges, burrs, or other safety hazards?
[29 CFR 1910.29(a)(2)(v)] ☐ ☐ ☐

2. Is the maximum work level height less than or equal to four times the minimum or least base dimension of the mobile ladder stand?
[29 CFR 1910.29(a)(3)(i)] ☐ ☐ ☐

Note: Suitable outrigger frames may be used to achieve the required base dimension or other means used to guy or brace the unit against tipping.

3. Is the minimum step width for ladder stands 16 in (40.64 cm)?
   [29 CFR 1910.29(a)(3)(ii)]

4. Are the steps of ladder stands fabricated from slip-resistant treads?
   [29 CFR 1910.29(a)(3)(iv)]

5. Are at least two of the four casters equipped with a swivel lock to prevent movement?
   [29 CFR 1910.29(a)(4)(ii)]

6. Are steps of mobile ladder stands uniformly spaced?
   [29 CFR 1910.29(f)(3)]

7. Are steps of mobile ladder stands sloped, with a rise that is not less than 9 in (22.86 cm) and not more than 10 in (25.4 cm), and a depth of at least 7 in (17.78 cm)?
   [29 CFR 1910.29(f)(3)]

Note: The slope of the steps section shall be a minimum of 55 degrees and a maximum of 60 degrees measured from the horizontal.

8. Are mobile ladder stands with more than five steps equipped with handrails?
   [29 CFR 1910.29(f)(4)(i)]

9. Are the handrails at least 29 in (73.66 cm) high?
   [29 CFR 1910.29(f)(4)(ii)]

Note: Measurements must be taken vertically from the center of the steps.

10. Are all ladder stands with a work level of 10 ft (3.05 cm) or higher above the ground or floor equipped with a standard 4-in (10.16-cm) nominal toeboard?
    [29 CFR 1910.29(a)(3)(vi)]

10.5 Construction Site Ladders

This checklist covers regulations issued by the U.S. Department of Labor, Occupational Safety and Health Administration (OSHA) under the construction standards 29 CFR 1926.1050 to 1926.1060.

It applies to portable ladders used at temporary worksites associated with construction, alteration, demolition, or repair work, including painting and decorating.

The regulations cited apply only to private employers and their employees, unless adopted by a state agency and applied to other groups such as public employees. Definitions of terms in bold type are provided at the end of the checklist.

LADDERS OR STAIRWAYS   YES   NO   N/A

1. Are ladders or stairways provided at all points of access that are elevated 19 in (48.26 cm) or more, and no ramp, runway, sloped embankment, or personnel hoist is provided?
[29 CFR 1926.1051(a)]   ☐   ☐   ☐

2. Does a competent person provide ladder training that teaches users how to recognize hazards and procedures for minimizing these hazards?
[29 CFR 1926.1060(a)]   ☐   ☐   ☐

3. Is ladder retraining provided when necessary?
[29 CFR 1926.1060(b)]   ☐   ☐   ☐

4. Can ladders support the load they are expected to carry?
[29 CFR 1926.1053(a)(1)(i),(ii),and(iii)]   ☐   ☐   ☐

5. Are ladder rungs, cleats, and steps parallel, level, and uniformly spaced when the ladder is in in position for use?
[29 CFR 1926.1053(a)(2)]   ☐   ☐   ☐

6. Are rungs, cleats, and steps of portable ladders (other than step stools and extension trestle ladders) spaced at least 10 in (25.4 cm) but not more than 14 in (35.56 cm) apart, as measured between center lines of the rungs, cleats, and steps?
[29 CFR 1926.1053(a)(3)(i)]   ☐   ☐   ☐

7. Are rungs, cleats, and steps of step stools at least 8 in (20.32 cm) but not more than 12 in (30.48 cm) apart, as measured between center lines of the rungs, cleats, and steps?
[29 CFR 1926.1053(a)(3)(ii)]   ☐   ☐   ☐

8. Are rungs, cleats, and steps of the base section of extension trestle ladders at least 8 in (20.32 cm) but not more than 18 in (45.72 cm) apart, as measured between center lines of the rungs, cleats, and steps?
[29 CFR 1926.1053(a)(3)(iii)]   ☐   ☐   ☐

9. Are rungs, cleats, and steps of the extension section of extension trestle ladders at least 6 in (15.24 cm) but not more than 12 in. (30.48 cm) apart as measured between center lines of the rungs, cleats, and steps? [29 CFR 1926.1053(a)(3)(iii)] ☐ ☐ ☐

10. Is clear distance between side rails for all portable ladders at least 11.5 in (29.21 cm)? [29 CFR 1926.1053(a)(4)(ii)] ☐ ☐ ☐

11. Are the rungs and steps of portable metal ladders corrugated, knurled, dimpled, coated with skid-resistant material, or otherwise treated to minimize slipping? [29 CFR 1926.1053(a)(6)(ii)] ☐ ☐ ☐

12. Are workers PROHIBITED from tying or fastening ladders together to provide longer sections (unless they are designed for such use)? [29 CFR 1926.1053(a)(7)] ☐ ☐ ☐

13. Is a metal spreader or locking device provided on each stepladder to hold the front and back sections in an open position when the ladder is being used? [29 CFR 1926.1053(a)(8)] ☐ ☐ ☐

14. Are ladder surfaced to prevent injury from punctures or lacerations, and to prevent snagging of clothing? [29 CFR 1926.1053(a)(11)] ☐ ☐ ☐

15. Are workers PROHIBITED to coat wood ladders with any opaque covering, except for identification or warning labels that are placed on only one face of a side rail? [29 CFR 1926.1053(a)(12)] ☐ ☐ ☐

16. Do portable ladders extend at least 3 ft (91.44 cm) above the upper landing surface for which the ladder is used to gain access? [29 CFR 1926.1053(b)(1)] ☐ ☐ ☐

Note: As an alternative, secure the ladder at its top to a rigid support that will not deflect. Use a grasping device (such as a grab rail) to mount and dismount the ladder. The extension should never be such that the ladder deflection under load would, by itself, cause the ladder to slip off its support.

17. Are ladders maintained free of oil, grease, and other slipping hazards?
    [29 CFR 1926.1053(b)(2)]
    ☐ ☐ ☐

18. Are ladders loaded at or below the maximum intended load for which they were built, or at or below the manufacturer's rated capacity?
    [29 CFR 1926.1053(b)(3)]
    ☐ ☐ ☐

19. Are ladders only used for the purpose for which they were designed?
    [29 CFR 1926.1053(b)(4)]
    ☐ ☐ ☐

20. Are non-self-supporting ladders used at an angle such that the horizontal distance from the top support to the foot of the ladder is approximately one-fourth of the working length of the ladder (the distance along the ladder between the foot and the top support)?
    [29 CFR 1926.1053(b)(5)(i)]
    ☐ ☐ ☐

21. Are ladders used only on stable and level surfaces, unless they are secured to prevent displacement?
    [29 CFR 1926.1053(b)(6)]
    ☐ ☐ ☐

22. Are ladders used on slippery surfaces ONLY when they are secured or provided with slip-resistant feet to prevent displacement?
    [29 CFR 1926.1053(b)(7)]
    ☐ ☐ ☐

Note: Do not use slip-resistant feet as a substitute for care in placing, lashing, or holding a ladder on surfaces such as flat metal or concrete that cannot be prevented from becoming slippery.

23. Are ladders secured to prevent displacement, especially in busy, high-traffic areas?
    [29 CFR 1926.1053(b)(8)]
    ☐ ☐ ☐

Note: As an alternative, a barricade may be used to keep the activities or traffic away from the ladder.

24. Is the area around the top and bottom of ladders kept clear?
    [29 CFR 1926.1053(b)(9)]
    ☐ ☐ ☐

25. Is the top of a non-self-supporting ladder placed with the two rails supported equally, unless it has a single support attachment?
    [29 CFR 1926.1053(b)(10)]
    ☐ ☐ ☐

26. Are workers PROHIBITED from moving, shifting, or ☐ ☐ ☐
    extending ladders while they are occupied?
    [29 CFR 1926.1053(b)(11)]
27. Do ladders have nonconductive side rails if they are ☐ ☐ ☐
    used where they could contact exposed energized
    electric equipment?
    [29 CFR 1926.1053(b)(12)]
28. Are workers PROHIBITED from standing on the top, ☐ ☐ ☐
    or is the top step of a stepladder prohibited?
    [29 CFR 1926.1053(b)(13)]
29. Are workers PROHIBITED from climbing on the ☐ ☐ ☐
    cross-bracing on the rear section of stepladders?
    [29 CFR 1926.1053(b)(14)]

Note: This is allowed ONLY if the ladder is designed and provided with steps for climbing on both front and rear sections.

30. Are ladders inspected periodically by a competent ☐ ☐ ☐
    person and after any incident that could affect their
    safe use?
    [29 CFR 1926.1053(b)(15)]
31. Are portable ladders with structural defects
    - Immediately marked in a manner that readily ☐ ☐ ☐
      identifies them as defective?
    - Tagged with DO NOT USE or similar language? ☐ ☐ ☐
    - Withdrawn from service until repaired? ☐ ☐ ☐
      [29 CFR 1926.1053(b)(16)]

Note: Structural defects include broken or missing rails, corroded components, or other faulty or defective components.

32. Does a ladder that is repaired meet its original ☐ ☐ ☐
    design criteria, before it is returned to use?
    [29 CFR 1926.1053(b)(18)]
33. Do all workers face the ladder when moving up or ☐ ☐ ☐
    down the ladder?
    [29 CFR 1926.1053(b)(20)]
34. Do all students and employees use at least one hand ☐ ☐ ☐
    to grasp the ladder when moving up or down the
    ladder?
    [29 CFR 1926.1053(b)(21)]

35. Are workers PROHIBITED from carrying any object ☐ ☐ ☐
or load that could cause a person to lose balance
and fall?
[29 CFR 1926.1053(b)(22)]

Source: NIOSH—National Institute for Occupational Safety & Health, Cincinnati, OH.

PORTABLE LADDERS

This checklist covers regulations issued by the U.S. Department of Labor, Occupational Safety and Health Administration (OSHA) under the construction standards 29 CFR 1926.1050 to 1926.1060.

It applies to portable ladders used at temporary worksites associated with construction, alteration, demolition, or repair work, including painting and decorating.

The regulations cited apply only to private employers and their employees, unless adopted by a state agency and applied to other groups such as public employees.

	YES	NO	N/A
1. Are ladders or stairways provided at all points of access that are elevated 19 in (48.26 cm) or more, and no ramp, runway, sloped embankment, or personnel hoist is provided? [29 CFR 1926.1051(a)]	☐	☐	☐
2. Does a competent person provide ladder training that teaches users how to recognize hazards and procedures for minimizing these hazards? [29 CFR 1926.1060(a)]	☐	☐	☐
3. Is ladder retraining provided when necessary? [29 CFR 1926.1060(b)]	☐	☐	☐
4. Can ladders support the load they are expected to carry? [29 CFR 1926.1053(a)(1)(i),(ii),and(iii)]	☐	☐	☐
5. Are ladder rungs, cleats, and steps parallel, level, and uniformly spaced when the ladder is in position for use? [29 CFR 1926.1053(a)(2)]	☐	☐	☐
6. Are rungs, cleats, and steps of portable ladders (other than step stools and extension trestle ladders) spaced at least 10 in (25.4 cm) but not more than 14 in (35.56 cm) apart (as measured between center lines of the rungs, cleats, and steps)? [29 CFR 1926.1053(a)(3)(i)]	☐	☐	☐

7. Are rungs, cleats, and steps of step stools at least 8 in (20.32 cm) but not more than 12 in (30.5 cm) apart (as measured between center lines of the rungs, cleats, and steps)?
[29 CFR 1926.1053(a)(3)(ii)] ☐ ☐ ☐

8. Are rungs, cleats, and steps of the base section of extension trestle ladders at least 8 in (20.32 cm) but not more than 18 in (45.72 cm) apart (as measured between center lines of the rungs, cleats, and steps)?
[29 CFR 1926.1053(a)(3)(iii)] ☐ ☐ ☐

9. Are rungs, cleats, and steps of the extension section of extension trestle ladders at least 6 in (20.32 cm) but not more than 12 in (30.5 cm) apart (as measured between center lines of the rungs, cleats, and steps)?
[29 CFR 1926.1053(a)(3)(iii)] ☐ ☐ ☐

10. Is the clear distance between side rails for all portable ladders at least 11.5 in (29.21 cm)?
[29 CFR 1926.1053(a)(4)(ii)] ☐ ☐ ☐

11. Are the rungs and steps of portable metal ladders corrugated, knurled, dimpled, coated with skid-resistant material, or otherwise treated to minimize slipping?
[29 CFR 1926.1053(a)(6)(ii)] ☐ ☐ ☐

12. Are ladders prohibited from being tied or fastened together to provide longer sections (unless they are designed for such use)?
[29 CFR 1926.1053(a)(7)] ☐ ☐ ☐

13. Is a metal spreader or locking device provided on each stepladder to hold the front and back sections in an open position when the ladder is being used?
[29 CFR 1926.1053(a)(8)] ☐ ☐ ☐

14. Are ladder surfaced to prevent injury from punctures or lacerations, and to prevent snagging of clothing?
[29 CFR 1926.1053(a)(11)] ☐ ☐ ☐

15. Are workers PROHIBITED to coat wood ladders with any opaque covering, except for identification or warning labels that are placed on only one face of a side rail?
[29 CFR 1926.1053(a)(12)] ☐ ☐ ☐

16. Do portable ladders extend at least 3 ft above the upper landing surface for which the ladder is used to gain access?
[29 CFR 1926.1053(b)(1)]

☐ ☐ ☐

Note: As an alternative, secure the ladder at its top to a rigid support that will not deflect. Use a grasping device (such as a grab rail) to mount and dismount the ladder. The extension should never be such that the ladder deflection under load would, by itself, cause the ladder to slip off its support.

17. Are ladders maintained free of oil, grease, and other slipping hazards?
[29 CFR 1926.1053(b)(2)]

☐ ☐ ☐

18. Are ladders loaded at or below the maximum intended load for which they were built, or at or below the manufacturer's rated capacity?
[29 CFR 1926.1053(b)(3)]

☐ ☐ ☐

19. Are ladders only used for the purpose for which they were designed?
[29 CFR 1926.1053(b)(4)]

☐ ☐ ☐

20. Are non-self-supporting ladders used at an angle such that the horizontal distance from the top support to the foot of the ladder is approximately one-fourth of the working length of the ladder (the distance along the ladder between the foot and the top support)?
[29 CFR 1926.1053(b)(5)(i)]

☐ ☐ ☐

21. Are ladders used ONLY on stable, level surfaces, unless secured to prevent displacement?
[29 CFR 1926.1053(b)(6)]

☐ ☐ ☐

22. Are ladders used on slippery surfaces ONLY when they are secured or provided with slip-resistant feet to prevent displacement?
[29 CFR 1926.1053(b)(7)]

☐ ☐ ☐

Note: Do not use slip-resistant feet as a substitute for care in placing, lashing, or holding a ladder on surfaces such as flat metal or concrete that cannot be prevented from becoming slippery.

23. Are ladders secured to prevent displacement, especially in busy, high-traffic areas?
[29 CFR 1926.1053(b)(8)]

☐ ☐ ☐

Note: As an alternative, a barricade may be used to keep the activities or traffic away from the ladder.

24. Is the area around the top and bottom of ladders kept clear?
[29 CFR 1926.1053(b)(9)] ☐ ☐ ☐

25. Is the top of a non-self-supporting ladder placed with the two rails supported equally, unless it has a single support attachment?
[29 CFR 1926.1053(b)(10)] ☐ ☐ ☐

26. Are workers PROHIBITED from moving, shifting, or extending ladders while they are occupied?
[29 CFR 1926.1053(b)(11)] ☐ ☐ ☐

27. Do ladders have nonconductive side rails if they are used where they could contact exposed energized electric equipment?
[29 CFR 1926.1053(b)(12)] ☐ ☐ ☐

28. Are workers PROHIBITED from standing on the top or top step of a stepladder?
[29 CFR 1926.1053(b)(13)] ☐ ☐ ☐

29. Are workers PROHIBITED from climbing on the cross-bracing on the rear section of stepladders?
[29 CFR 1926.1053(b)(14)] ☐ ☐ ☐

Note: This is allowed ONLY if the ladder is designed and provided with steps for climbing on both front and rear sections.

30. Are ladders inspected periodically by a competent person and after any incident that could affect their safe use?
[29 CFR 1926.1053(b)(15)] ☐ ☐ ☐

31. Are portable ladders with structural defects
    - Immediately marked in a manner that readily identifies them as defective? ☐ ☐ ☐
    - Tagged with DO NOT USE or similar language? ☐ ☐ ☐
    - Withdrawn from service until repaired? ☐ ☐ ☐
[29 CFR 1926.1053(b)(16)]

Note: Structural defects include broken or missing rails, corroded components, or other faulty or defective components.

32. Does a ladder that is repaired meet its original design criteria, before it is returned to use?
[29 CFR 1926.1053(b)(18)]   ☐ ☐ ☐

33. Do all workers face the ladder when moving up or down the ladder?
[29 CFR 1926.1053(b)(20)]   ☐ ☐ ☐

34. Do all workers use at least one hand to grasp the ladder when moving up or down ladder?
[29 CFR 1926.1053(b)(21)]   ☐ ☐ ☐

35. Are workers PROHIBITED from carrying any object or load that could cause a person to lose balance and fall?
[29 CFR 1926.1053(b)(22)]   ☐ ☐ ☐

Standards (consensus)

ANSI consensus standards on portable ladders include

ANSI A14.1-2007	Safety Requirements for Wood Ladders
ANSI A14.2-2007	Safety Requirements for Portable Metal Ladders
ANSI A14.4-2009	Safety Requirements for Job-Made Ladders
ANSI A14.5-2000	Safety-Reinforced Plastic Ladders
ANSI A14.7-2006	Safety Requirements for Mobile Ladders

These standards detail specifications on the various materials, construction requirements, test requirements, usage guidelines, and labeling/marking requirements for portable ladders.

ANSI recommends various species of wood that are suitable for ladders. Physical characteristics such as grain, knot, pitch, and compression must be controlled in the construction of ladders.

Reinforced plastic ladders must use fully cured, commercial-grade thermosetting polyester resin with glass fiber reinforcement. The type of material to be used is determined by the environment that the finished ladder will encounter (electrical hazards, temperature extremes, corrosion, outdoor weathering, etc.). Metal ladders do not have material guidelines.

Size categories vary for wood, metal and reinforced plastic materials, ladder types, and ladder designs (stepladder, extension ladder, platform ladder, etc.).

Ladder type	Maximum length	Special requirements
Single section	30 ft	The minimum width between side rails of a straight ladder or any section of an extension ladder shall be 12 in. The length of single ladders or individual sections of ladders shall not exceed 30 ft.
Extension ladders	48 ft	Two-section ladders shall not exceed 48 ft in length.
Greater than two-section	60 ft	Greater than two-section ladders shall not exceed 60 ft in length. Overlap stops required.
Stepladders	20 ft	Insulating, nonslip pads at bottom of rails. Must have locking device to hold ladder sections open.
Platform ladder	20 ft	None
Trestle ladder/extensions	20 ft	None

Ladder type	Duty rating	Description
Type 1 AA	375 lb	Extra heavy-duty industrial ladder
Type 1 A	300 lb	Heavy-duty industrial ladder
Type 1	250 lb	Heavy-duty industrial ladder
Type 2	225 lb	Medium-duty industrial ladder
Type 3	200 lb	Light-duty industrial ladder

Warning: Ladders MUST be marked with ladder size, type, maximum length, number of sections (if appropriate), highest standing level, total length of sections (if applicable), model number, manufacturer's name, manufacturer's location, and date of manufacture.

Usage guidelines and other warning statements MUST also be placed on the ladders in specific locations depending on ladder type.

10.6 Fixed Stairways and Ramps

Slips, trips, and falls on stairways are a major source of injuries and fatalities among construction workers. To help eliminate these hazards:

- Stairway treads and walkways MUST be free of dangerous objects, debris, and materials.
- Slippery conditions on stairways and walkways MUST be corrected immediately.

- Make sure that treads cover the entire step and landing.
- Stairways having four or more risers or rising more than 30 in MUST have at least one handrail.

OSHA regulations apply to interior and exterior stairs around machinery, tanks, and other equipment, and stairs leading to or from floors, platforms, or pits.

They DO NOT apply to stairs used for fire exit purposes. It also does not address fixed stairs associated with construction sites. Consult the OSHA regulations 29 CFR 1926.1051 and 1926.1052 for construction site requirements.

The regulations cited apply only to private employers and their employees, unless adopted by a state agency and applied to other groups such as public employees.

FIXED STAIRS   YES   NO   N/A

1. Are fixed stairs (rather than ladders or other means of access) provided where access to elevation is necessary on a daily or regular basis?
   [29 CFR 1910.24(b)]

2. Do fixed stairs have a minimum width of 22 in (55.88 cm)?
   [29 CFR 1910.24(d)]

3. Are fixed stairs installed at angles to the horizontal between 30 and 50 degrees?
   [29 CFR 1910.24(e)]

4. Are all treads reasonably slip-resistant with the front protruding edge of the tread of a nonslip finish?
   [29 CFR 1910.24(f)]

5. Do fixed stairs have a uniform rise height and tread width throughout the flight of stairs?
   [29 CFR 1910.24(f)]

6. Are stairway landing platforms no less than the width of the stairway and a minimum of 30 in (76.2 cm) long measured in the direction of travel?
   [29 CFR 1910.24(g)]

7. Are standard railings provided on all open sides of exposed stairways and stair platforms? (See checklist Guarding Floors, Stairs, and Other Openings)
   [29 CFR 1910.24(h)])

8. Is a vertical clearance above the stair tread to an overhead obstruction that is at least 7 ft (2.14 m) measured from the leading edge of the tread? [29 CFR 1910.24(i)] ☐ ☐ ☐

Stairs and Stairways

9. Are standard stair rails or handrails present on all stairways having at least four risers? ☐ ☐ ☐

10. Do stairs have at least 6.5 ft (1.98 m) of overhead clearance? ☐ ☐ ☐

11. Do stairs have landing platforms not less than 30 in (76.20 cm) in the direction of travel and extend 22 in (55.88 cm) in width at every 12 ft (3.6576 m) or less of vertical rise? ☐ ☐ ☐

12. Do stairs angle no more than 50 degrees and no less than 30 degrees? ☐ ☐ ☐

13. Are stairs of hollow pan-type treads and landings filled to the top edge of the pan with solid material? ☐ ☐ ☐

14. Are step risers on stairs uniform from top to bottom, with no riser spacing greater than 9.5 in (24.13 cm)? ☐ ☐ ☐

15. Are steps on stairs and stairways designed or provided with a surface that renders them slip-resistant? ☐ ☐ ☐

16. Are stairway handrails located between 30 in (76.2 cm) and 42 in (106.68 cm) above the leading edge of stair treads? ☐ ☐ ☐

17. Do stairway handrails have at least 3 in (76.2 cm) of clearance between the handrails and wall or surface they are mounted on? ☐ ☐ ☐

18. Where doors or gates open directly on a stairway, is a platform provided so the swing of the door does not reduce the width of the platform to less than 21 in (53.34 cm)? ☐ ☐ ☐

19. Are stairway handrails capable of withstanding a load of 200 lb (90.7 kg) applied within 2 in (5.08 cm) of the top edge in any downward or outward direction? ☐ ☐ ☐

20. Where stairs or stairways exit directly into any area where vehicles may be operated, are adequate barriers and warnings provided to prevent employees from stepping into the path of traffic? ☐ ☐ ☐
21. Do stairway landings have a dimension measured in the direction of travel at least equal to the width of the stairway? ☐ ☐ ☐
22. Is the vertical distance between stairway landings limited to 12 ft (3.66 m) or less? ☐ ☐ ☐

Ramps

23. Do ramps and walkways—6 ft (1.8 m) or more above lower levels—have guardrail systems that comply with OSHA 1926 subpart M—Fall Protection? ☐ ☐ ☐
24. Are NO ramps or walkways inclined more than a slope of one (1) vertical to three (3) horizontal (20 degrees above the horizon)? ☐ ☐ ☐

Note: If the slope of a ramp or a walkway is steeper than one (1) vertical in eight (8) horizontal, OSHA requires that the ramp or walkway shall have cleats not more than fourteen inches (35 cm) apart—that are securely fastened to the planks to provide footing.

10.7 Walkways and Elevated Platforms

GENERAL AREAS AND WALKWAYS

	YES	NO	N/A
1. Are all worksites (i.e., offices, store rooms, shops) clean and orderly?	☐	☐	☐
2. Are work surfaces kept dry or covered with nonslip materials?	☐	☐	☐
3. Are all aisles and passageways at least 22 in (55.88 cm) wide?	☐	☐	☐
4. Are aisles and passageways kept clear and marked as appropriate?	☐	☐	☐
5. Are holes in the floor, sidewalk, or other walking surface repaired properly, covered, or otherwise made safe?	☐	☐	☐
6. Is there safe clearance for walking in aisles where motorized or mechanical handling equipment is operating?	☐	☐	☐

7. Are materials or equipment stored in such a way so that sharp objects DO NOT protrude and cannot obstruct the walkway? ☐ ☐ ☐

8. Are spilled materials cleaned up immediately? ☐ ☐ ☐

9. Are changes of direction or elevations readily identifiable? ☐ ☐ ☐

10. Are aisles or walkways that pass near moving or operating machinery, welding operations, or similar operations arranged so employees will not be subjected to potential hazards? ☐ ☐ ☐

11. Is there adequate headroom, at least 6.5 ft (1.98 m), provided for the entire length of any aisle or walkway? ☐ ☐ ☐

12. Are approved 42-in (106.68-cm) high guardrails provided wherever aisles, landings, or other walkway surfaces are elevated more than 30 in (76.20 cm) above the ground? ☐ ☐ ☐

13. Are bridges provided over conveyors and similar hazards? ☐ ☐ ☐

ELEVATED SURFACES   YES   NO   N/A

1. Are signs posted, when appropriate, showing the elevated surface (floor) load capacity? ☐ ☐ ☐

2. Are surfaces, elevated more than 30 in (76.20 cm), provided with standard guardrails? ☐ ☐ ☐

3. Are all elevated surfaces beneath, which people or machinery could be exposed to falling objects, provided with standard 4-in (10.16-cm) toeboards? ☐ ☐ ☐

4. Is a permanent means of access and egress provided to elevated storage and work surfaces? ☐ ☐ ☐

5. Is required headroom provided where necessary? ☐ ☐ ☐

6. Is material on elevated surfaces piled, stacked, or racked in a manner to prevent it from tipping, falling, collapsing, rolling, or spreading? ☐ ☐ ☐

7. Are dock boards or bridge plates used when transferring materials between docks and trucks or railcars? ☐ ☐ ☐

8. When in use, are dock boards or bridge plates secured in place? ☐ ☐ ☐

FLOOR AND WALL OPENINGS   YES   NO   N/A

1. Are floor openings (holes) guarded by a cover, a   ☐   ☐   ☐
   guardrail, or equivalent on all sides (except at
   stairways or ladder entrances)?

2. Are toeboards installed around the edges of   ☐   ☐   ☐
   permanent floor openings where persons may
   pass below the opening?

3. Are skylight screens of such construction and   ☐   ☐   ☐
   mounting that they will withstand a load of at
   least 200 lb (90.7 kg)?

4. Is the glass in windows, doors, glass walls, etc., subject   ☐   ☐   ☐
   to possible human impact and of sufficient thickness
   and type for the condition of use?

5. Are grates or similar type covers over floor openings,   ☐   ☐   ☐
   such as floor drains, of such design that foot traffic or
   rolling equipment will not be caught or impeded by
   the grate spacing?

6. Are unused portions of service pits and pits not in   ☐   ☐   ☐
   use either covered or protected by guardrails or
   equivalent?

7. Are manhole covers, trench covers, and similar covers,   ☐   ☐   ☐
   and their supports designed to carry a truck rear axle
   load of at least 20,000 lb (9072 kg) when located in
   roadways and subject to vehicle traffic?

8. Are floor or wall openings in fire-resistant construction ☐   ☐   ☐
   provided with doors or covers compatible with the
   fire rating of the structure and provided with a
   self-closing feature when appropriate?

10.8 Guarding Roofs, Floors, Stairs, and Other Openings

This checklist covers regulations issued by the U.S. Department of Labor, Occupational Safety and Health Administration (OSHA) under the general industry standard 29 CFR 1910.23.

It applies to spaces having permanent and temporary floor holes and openings greater than 1 in (2.54 cm) in its least dimension, such as floor drains, manholes, hatchways, ladder openings, or pits, and raised open-sided floors, platforms, runways, or storage areas.

Ladders, Stairways, and Ramps 269

	YES	NO	N/A

1. Is every skylight floor opening and hole guarded by a standard skylight screen or a fixed standard railing on all exposed sides?
   [29 CFR 1910.23(a)(4)] ☐ ☐ ☐

2. Are all floor openings to stairways, ladderways, hatchways, chutes, or manholes guarded by a standard railing and toeboards—on all sides except at the entrance—or other protective cover?
   [29 CFR 1910.23(a)(1), (2), (3), (5), and (6)] ☐ ☐ ☐

3. Is every temporary floor opening guarded by a standard railing or constantly attended by someone?
   [29 CFR 1910.23(a)(7)] ☐ ☐ ☐

4. Is every floor hole into which a person could fall guarded by either a standard railing and toe board or floor hole cover?
   [29 CFR 1910.23(a)(8)] ☐ ☐ ☐

5. Is every floor hole into which a person could not fall (because of fixed machinery, equipment, or walls) protected by a cover that leaves no openings of more than 1 in (2.54 cm) width?
   [29 CFR 1910.23(a)(9)] ☐ ☐ ☐

Note: The cover MUST be securely held in place to prevent tools or materials from falling through.

6. Does a platform allow an effective width of at least 20 in (50.8 cm) where doors or gates open directly onto a stairway, when the door swings open?
   [29 CFR 1910.23(a)(10)] ☐ ☐ ☐

7. Is every open-sided floor or platform that is 4 ft (1.22 cm) or more above the adjacent floor ground level guarded by a standard railing on all open sides?
   [29 CFR 1910.23(c)(1)] ☐ ☐ ☐

8. Is every runway guarded by a standard railing on all open sides that are 4 ft (1.22 cm) or more above the floor or ground level?
   [29 CFR 1910.23(c)(2)] ☐ ☐ ☐

9. Regardless of height, are all open-sided floors, walkways, platforms, or runways guarded by a standard railing and toeboard if they are above or adjacent to any dangerous equipment or operation?
   [29 CFR 1910.23(c)(3)] ☐ ☐ ☐

10. Is a railing provided with a toeboard wherever, beneath the open sides,
    - Persons can pass? ☐ ☐ ☐
    - Machinery is moving? ☐ ☐ ☐
    - Equipment could create a hazard of falling materials? ☐ ☐ ☐
    [29 CFR 1910.23(c)(1)]

11. Is every wall opening from which the drop is more than 4 ft (1.22 m) guarded with a standard railing or other barrier? ☐ ☐ ☐
    [29 CFR 1910.23(b)(1), (2) and (4)]

12. Is every window wall opening guarded by slats, grill work, or standard railing if
    - It is at a stairway landing, floor, platform, or balcony from which the drop is more than 4 ft (1.22 m)? ☐ ☐ ☐
    - The bottom of the opening is less than 3 ft (91.44 cm) above the platform or landing? ☐ ☐ ☐
    [29 CFR 1910.23(b)(3)]

13. Is every flight of stairs with four or more risers equipped with standard stair railings or standard handrails as specified below? ☐ ☐ ☐
    [29 CFR 1910.23(d)(1)]

14. Where standard railings are provided, do they meet the following specifications:
    - On stairways, less than 44 in (1.12 m) wide
        with both sides enclosed—at least one handrail is required, preferably on the right-hand side descending? ☐ ☐ ☐
        with one open side—at least one stair railing must be on the open side? ☐ ☐ ☐
        with both sides open—one stair railing is required on each side? ☐ ☐ ☐
    - On stairways, 44 to 88 in (1.12–2.24 m) wide
        one handrail on each enclosed side? ☐ ☐ ☐
        one stair railing on each open side is required? ☐ ☐ ☐
    - On stairways, 88 in (2.48 m) or more wide
        one handrail on each enclosed side? ☐ ☐ ☐
        one stair railing on each open side? ☐ ☐ ☐
        one intermediate stair railing located approximately midway of the width? ☐ ☐ ☐
    [29 CFR 1910.23(e)(1)]

Note: The rail MUST consist of a top rail at a height of 42 in (1.07 cm) and a midrail at approximately 21 in (53.34 cm), and the top rail MUST be smooth surfaced throughout the length of the railing.

15. If wooden railings are used for guardrails, are the posts at least 2 by 4 in (5.08 × 10.16 cm) and spaced less than 6 ft (1.83 m) apart?
[29 CFR 1910.23(e)(3)(I)] ☐ ☐ ☐

Note: The top rail and intermediate rails must also be at least 2 by 4 in (5.08 × 10.16 cm) stock.

16. If pipe railings are used, are posts and top and intermediate rails at least 1.5 in (3.81 cm) nominal diameter with posts spaced less than 8 ft (2.44 m) on centers?
[29 CFR 1910.23(e)(3)(ii)] ☐ ☐ ☐

17. If structural steel is used for guardrails, are the posts and top and intermediate rails, at least 2 × 2 × 3/8 in (5.08 × 5.08 cm × 9.53 mm) angle irons, or other metal shapes of equivalent bending strength with posts spaced not more than 8 ft (2.44 m) on centers?
[29 CFR 1910.23(e)(3)(iii)] ☐ ☐ ☐

18. Is the guardrail anchored and of such construction that it is capable of withstanding a load of at least 200 lb (90.72 kg) applied in any direction at any point on the top rail?
[29 CFR 1910.23(e)(3)(iv)] ☐ ☐ ☐

19. Are standard toeboards at least 4 in (10.16 cm) in height provided at the floor of the guardrail?
[29 CFR 1910.23(e)(4)] ☐ ☐ ☐

20. Are handrails constructed so that they can be easily grasped (i.e., rounded)?
[29 CFR 1910.23(e)(5)] ☐ ☐ ☐

21. Are all handrails and railings provided with a clearance of at least 3 in (7.62 cm) between the handrail or railing and any other object?
[29 CFR 1910.23(e)(6)] ☐ ☐ ☐

Note: A distance less than this would make it difficult to get a good grasp in an emergency.

22. Are skylight screens constructed so that they are capable of withstanding a load of at least 200 lb (90.72 kg) applied perpendicularly to any area on the screen?
[29 CFR 1910.23(e)(8)]

☐ ☐ ☐

Note: Sometimes people get on the roof and fall through skylight screens that are not designed to prevent this type of fall.

23. Are wall opening barriers (rails, rollers, picket fences, half doors) constructed and mounted so that the barrier is capable of withstanding a load of at least 200 lb (90.72 kg) applied in any direction (except upward) at any point on top rail or corresponding member?
[29 CFR 1910.23(e)(9)]

☐ ☐ ☐

Source: NIOSH—National Institute for Occupational Safety and Health, Cincinnati, OH.

10.9 Standards (Consensus)

ANSI A1264.1 1995 (R2002)	Safety Requirements for Workplace Floor and Wall Openings
ANSI/ASSE A1264.1-2007	Safety Requirements for Workplace Walking/Working Surfaces and Their Access; Workplace Floor, Wall and Roof Openings Stairs and Guardrail Systems

Chapter 11

Tools and Machinery

11.1 Hand and Portable Power Tools

Construction workers use many hand tools, categorized as

- Non-powered—Including hammers, screwdrivers, pliers, wenches, adzes, axes, crow bars, and pry bars, screw drivers, wrenches
- Power tools (classified by power source)—Including electric, pneumatic, liquid fuel, hydraulic, and powder-actuated

Although tools are manufactured with safety for the user in mind, they pose potential hazards. If workers use hand tools over and over every day, they can injure their hands, wrists, or arms. They can be injured if they must hold on tight for a long time, or keep twisting a handle. They are subject to possible carpal syndrome, trigger finger, white finger, tendonitis, and other painful problems.

Hazards result from misuse and improper maintenance. The responsibility for controlling hazardous conditions on construction sites or in industrial facilities lies on both employers and employees.

- Employer is responsible for the safe condition of tools and equipment used by employees.
- Employee is responsible for proper use and maintenance of the equipment.

Employers MUST teach employees that sharps (saw blades, knives) and other tools be directed away from aisles and other employees working nearby. Knives and scissors must be sharp; however, dull tools can be more hazardous than sharp ones.

Floors MUST be kept as clean and dry as possible to prevent accidental slips with or around dangerous hand tools.

Only spark-resistant tools made from brass, plastic, aluminum, or wood should be used around flammable substances.

REGULATIONS	YES	NO	N/A
1. Are all portable hand or power tools maintained in a safe condition? [29 CFR 1926.300(a)]	☐	☐	☐
2. If compressed air is used for cleaning purposes, is it used at pressures less than 30 psi (2.11 kg/cm^2) and only with effective chip guarding and personal protective equipment (PPE)? [29 CFR 1910.242(b) and 1926.302(b)(4)]	☐	☐	☐
3. Are power tools equipped and used with guards whenever possible? [29 CFR 1926.300(b)(1)]	☐	☐	☐
4. Are all belts, gears, shafts, pulleys, sprockets, spindles, drums, flywheels, chains, or other reciprocating, rotating, or moving parts of equipment guarded if operator is exposed to contact or if they otherwise create a hazard? [29 CFR 1926.300(b)(2)]	☐	☐	☐
5. Is all necessary personal protective equipment provided whenever the use of hand and power tools could create falling, flying, or splashing debris, or harmful dusts, fumes, mists, vapors, or gases? [29 CFR 1926.300(c)]	☐	☐	☐
6. If workers use their own tools and equipment, are the tools and equipment subject to the same safety requirements as supplied tools and equipment? [29 CFR 1910.242(a)]	☐	☐	☐
7. Are all chain saws, percussion tools, and handheld powered circular saws [with blades greater than 2 in (5.08 cm) in diameter] equipped with a constant pressure switch that shuts off power when released? [29 CFR 1910.243(a)(2)(i); and 1926.300(d)(3)]	☐	☐	☐

8. Are all handheld powered drills; tappers; fastener ☐ ☐ ☐
drivers; horizontal, vertical, and angle grinders with
wheels greater than 2 in (5.08 cm) in diameter; disc
sanders with discs greater than 2 in (5.08 cm) in
diameter; belt sanders; reciprocating saws; saber,
scroll, jig saws (with blade shanks greater than a
nominal ¼ in (6.35 mm); and other similarly power
tools equipped with a constant pressure switch or
control?
[29 CFR 1910.243(a)(2)(ii) and 1926.300 (d)(1)-(3)]

Note: They may be equipped with a lock-on control provided the turnoff can be accomplished by a single motion by the same finger or fingers that turns it on. The construction standard requires a "momentary contact on-off control" instead of a constant pressure switch or control. This means that if the switch is pressed, the tool turns on and, if the switch is pressed again, the tool turns off.

9. Are all handheld powered platen sanders, grinders ☐ ☐ ☐
[with wheels 2 in (5.08 cm) diameter or less],
routers, planers, laminate trimmers, nibblers,
shears, scroll saws, and jig saws [with blade shanks
¼ in (6.35 mm) wide or less], equipped with a
positive "on-off" control?
[29 CFR 1910.243(a)(2)(iii) and 1926.300(d)(1)]

Note: A positive "on-off" control means a switch that you must push to turn the tool on and then push again to turn it off. Control switches as described in questions 7 and 8 may also be used.

10. On handheld power tools, is the operating control ☐ ☐ ☐
located so as to minimize the possibility of
accidental operation?
[29 CFR 1910.243(a)(2)(iv)]

Note: This requirement does not apply to concrete vibrators, concrete breakers, powered tampers, jackhammers, rock drills, garden appliances, household and kitchen appliances, personal care appliances, medical or dental equipment, or to fixed machinery.

11. Are all portable power-driven circular saws with ☐ ☐ ☐
blade diameter greater than 2 in (2.54 cm)
equipped with guards above and below the base
plate or shoe?
[29 CFR 1910.243(a)(1)(i) and 1926.304(d)]

12. Does the upper guard on a circular saw cover the saw to the depth of the teeth, except for the minimum arc required to permit the base to be tilted for bevel cuts?
    [29 CFR 1926.304(d)]

13. Does the lower guard on a circular saw cover the saw to the depth of the teeth, except for the minimum arc required to allow proper retraction and contact with the work?
    [29 CFR 1926.304(d)]

14. When a circular saw is removed from the material being sawed, does the lower guard automatically and instantly return to the covering position?
    [29 CFR 1926.304(d)]

15. Are belt sanding machines provided with guards at each nip point, where the sanding belt runs onto a pulley?
    [29 CFR 1910.243(a)(3)]

16. If a saw cracks, is it immediately removed from service?
    [29 CFR 1910.243(a)(4)]

17. Are all portable, electrically powered tools properly grounded or double insulated?
    [29 CFR 1910.243(a)(5) and 1926.302(a)(1)]

18. Are impact tools, such as drift pins, wedges, and chisels, kept free of mushroomed heads?
    [29 CFR 1926.301(c)]

19. Are the wooden handles of tools kept free of splinters or cracks and are they fixed tightly in the tool?
    [29 CFR 1926.301(d)]

20. Is it prohibited to lower or hoist a tool by its electric cord?
    [29 CFR 1926.302(a)(2)]

21. Do woodworking tools meet American National Standards Institute (ANSI) safety codes?
    [29 CFR 1926.304(f)]

Pneumatic Power Tools and Hoses

22. Are pneumatic power tools secured to the hose or whip by some positive means, so as to prevent the tool from being accidentally disconnected?
    [29 CFR 1926.302(b)(1)]

23. Are safety clips or retainers used on pneumatic impact (percussion) tools to prevent attachments from being accidentally expelled?
[29 CFR 1910.243(b)(1) and 1926.302(b)(2)]   ☐ ☐ ☐

24. Are all pneumatically driven nailers, staplers, and other similar equipment which have automatic fastener feeds and which operate at more than 100 psi (7.03 kg/cm^2) pressure at the tool equipped with a safety device on the nozzle to prevent the tool from ejecting fasteners, unless the muzzle is in contact with the work surface?
[29 CFR 1926.302(b)(3)]   ☐ ☐ ☐

25. Are all compressed air hoses and hose connections designed for the pressure and service to which they are subjected?
[29 CFR 1910.243(b)(2) and 1926.302(b)(5)]   ☐ ☐ ☐

26. Are workers PROHIBITED to lower or hoist tools by the hose?
[29 CFR 1926.302(b)(6)]   ☐ ☐ ☐

27. Do all hoses [exceeding 0.5 in (1.27 cm) inside diameter] have safety devices at the source of the supply or branch line, to reduce pressure in case of hose failure?
[29 CFR 1926.302(b)(7)]   ☐ ☐ ☐

28. Are airless spray guns of the type that atomize paints and fluids at high pressure—1000 psi (70.3 kg/cm^2)—equipped with automatic or visible manual safety devices that prevent accidental release of paint or fluid?
[29 CFR 1926.302(b)(8)]   ☐ ☐ ☐

Note: In lieu of the above, a diffuser nut which will prevent high-pressure, high-velocity release while the nozzle tip is removed, plus a nozzle tip guard which will prevent the tip from coming in contact with the operator (or other equivalent protection) shall be provided.

29. Are all fuel-powered tools stopped while being refueled, serviced, or maintained?
[29 CFR 1926.302(c)(1)]   ☐ ☐ ☐

30. Is all fuel transported, handled, and stored in accordance with applicable regulations?
[29 CFR 1926.302(c)(1)]   ☐ ☐ ☐

31. When fuel powered tools are used in enclosed spaces, are measures taken to prevent the buildup of toxic gases?
[29 CFR 1926.302(c)(2)]
☐ ☐ ☐

HAND AND POWER TOOLS YES NO N/A

1. Have all employees who may operate, or directly supervise those who operate, hand or portable power tools, or both, completed an authorized Hand and Power Tool Safety Training course? ☐ ☐ ☐

2. Have all employees who may use, or directly supervise those who use, powder-actuated tools completed an authorized Powder-Actuated Tools Safety Training course? ☐ ☐ ☐

3. Do the hazardous "nip points" and "entanglement points" on all saws, grinders, augers, power takeoffs (PTO's), and other power-operated tools and equipment have proper guarding that is required to be in place whenever used? ☐ ☐ ☐

4. Are electric supply cords and pneumatic and hydraulic hoses on power-operated tools checked regularly for deterioration or damage and repaired/replaced as required? ☐ ☐ ☐

5. Are all tools, machines, and equipment inspected on a regular basis and are maintenance logs kept updated? ☐ ☐ ☐

6. Is there a power shutoff switch within reach of the operator's position at each machine? ☐ ☐ ☐

7. Are stationary tools and equipment equipped with electromagnetic switches that prevent them from automatically restarting when power is restored after outages occur? ☐ ☐ ☐

8. Have all employees who operate, or directly supervise those who operate, mechanical, hydraulic, or pneumatic power presses completed an authorized Power Presses Safety training course? ☐ ☐ ☐

9. Have employees, and their supervisors, who use nylon slings, chains, and/or wire ropes in conjunction with hooks to lift and move materials completed an authorized Rigging Safety course? ☐ ☐ ☐

10. Are workers made aware of the hazards caused by faulty or improperly used hand tools? ☐ ☐ ☐
11. Are tools stored in dry, secure locations where they won't be tampered with? ☐ ☐ ☐
12. Are hand tools comfortable in the worker's hand—not too thick, too small, or too short? ☐ ☐ ☐

Note: Work with a hold handle tool can make some repetitive stress.

13. Do the hand tools conduct electricity or heat? ☐ ☐ ☐
14. Do the hand tools hurt users' hands when held tightly? ☐ ☐ ☐

Note: There should be NO sharp edges or finger grooves.

15. Do the hand tools have non-slip handles? ☐ ☐ ☐
    - Are handle surfaces made of soft materials—rubber or plastic? ☐ ☐ ☐

Note: Special plastic or rubber sleeves or custom grip kits are available to convert tool handles.

16. Is the handle long enough for a worker's whole hand (not just fingers) when considerable force is needed for the job? ☐ ☐ ☐
17. Are tools with thicker handles used when gloves are worn? ☐ ☐ ☐
18. Do some tools have a spring return that saves wear and tear on finger muscles, when reopening the tool after it is used? ☐ ☐ ☐
19. Are pistol-grip hand tools used where, for certain work, bent-angle or adjustable angle tools may strain worker wrists? ☐ ☐ ☐
20. Are hand tools kept sharp and in good condition, to reduce the force used on the tool and reduce stress on workers' hands and wrists? ☐ ☐ ☐
21. Are workers instructed
    - To NOT USE hand tools with bent wrists? ☐ ☐ ☐
    - To rest their hands during the day? ☐ ☐ ☐
    - To lay down or put the tool in a holster when it is NOT needed? ☐ ☐ ☐

22. Are hand tools comfortable and easy for workers to use? ☐ ☐ ☐
    - Are they appropriately designed to do the specific job? ☐ ☐ ☐

HAND TOOLS | YES | NO | N/A

1. Are hand tools and other equipment regularly inspected for safe condition? ☐ ☐ ☐
2. Are tool handles free of splits and cracks? ☐ ☐ ☐
3. Are handles wedged tightly in the heads of all tools? ☐ ☐ ☐
4. Are impact tools free of mushroomed heads? ☐ ☐ ☐
5. Are the heads of chisels or punches ground periodically to prevent mushrooming? ☐ ☐ ☐
6. Are cutting edges kept sharp so the tool will move smoothly without binding or skipping? ☐ ☐ ☐
7. Is sharpening, redressing, or repairing tools done properly, using tools suited to purpose? ☐ ☐ ☐
8. When compressed air is used for cleaning purposes, is nozzle pressure safely reduced to less than 30 psi (2.1 kg/cm^2)? ☐ ☐ ☐

POWER TOOLS | YES | NO | N/A

1. Are power saws and similar equipment provided with safety guards? ☐ ☐ ☐
2. Are tools used with the correct shield, guard, or attachments recommended by the manufacturer? ☐ ☐ ☐
3. Are abrasive wheel grinders provided with safety guards which cover the spindle ends, nut, and flange projections? ☐ ☐ ☐
4. Are work rests and tongue guards properly set? ☐ ☐ ☐
5. Are portable circular saws equipped with guards above and below the base or shoe? ☐ ☐ ☐
6. Are saw guards checked to ensure they are not wedged up, thereby leaving an unguarded lower portion of the blade? ☐ ☐ ☐
7. Are springs checked for wear? ☐ ☐ ☐
8. Are guards kept in place and in working order? ☐ ☐ ☐
9. Are rotating or moving parts of equipment guarded to prevent contact by employees? ☐ ☐ ☐
10. Do operators wear eye and face protection when grinding? ☐ ☐ ☐

11. Is the pneumatic hose line secured to the pneumatic tool? ☐ ☐ ☐
12. Are tool bits secured by use of safety clips or retainers to prevent accidental disconnecting? ☐ ☐ ☐
13. Are appropriate ground-fault circuit interrupters provided at the job site—OR is an assured equipment grounding conductor program established and implemented at the job site? ☐ ☐ ☐
14. Are electric tools double-insulated or properly grounded? ☐ ☐ ☐
15. Is eye and face protection used when driving hardened or tempered studs or nails? ☐ ☐ ☐
16. Are tools stored in a dry, secure location where they won't be tampered with? ☐ ☐ ☐

POWDER-ACTUATED TOOLS   YES   NO   N/A

1. Do all users of powder-actuated tools possess an "operator's card" to demonstrate that they have been trained? ☐ ☐ ☐
2. Are tools left unloaded until they are ready for immediate use? ☐ ☐ ☐
3. Are tools inspected each day for defects or obstructions prior to use? ☐ ☐ ☐
4. Are the following types of personal protective equipment used by operators and others in area of tool use:
   - Hard hats? ☐ ☐ ☐
   - Safety goggles? ☐ ☐ ☐
   - Safety shoes? ☐ ☐ ☐
   - Ear protection? ☐ ☐ ☐
   (Required when making fastening in confined areas—small room, tanks, vaults, or ship compartments.)

11.2 Welding and Cutting (Hot Work)

    YES   NO   N/A

1. Have employees who may perform, or supervise those who perform, welding or cutting activities completed an approved Welding and Cutting Safety course? ☐ ☐ ☐

2. Have employees who must perform hot work in areas other than approved, designated welding spaces completed a Hot Work Safety course? ☐ ☐ ☐
3. Are only authorized and trained personnel permitted to use welding, cutting, or brazing equipment? ☐ ☐ ☐
4. Does each operator have a copy of, and follow, the appropriate operating instructions? ☐ ☐ ☐
5. Are compressed-gas cylinders regularly examined for obvious signs of defects, deep rusting, or leakage? ☐ ☐ ☐
6. Is care used in handling and storage of cylinders, safety valves, relief valves, etc., to prevent damage? ☐ ☐ ☐
7. Are precautions taken to prevent the mixture of air or oxygen with flammable gases, except at a burner or in a standard torch? ☐ ☐ ☐
8. Are only approved apparatuses (torches, regulators, pressure-reducing valves, acetylene generators, manifolds) used? ☐ ☐ ☐
9. Are cylinders kept away from sources of heat and elevators, stairs, or gangways? ☐ ☐ ☐
10. Are employees prohibited from using cylinders as rollers or supports? ☐ ☐ ☐
11. Are empty cylinders appropriately marked, their valves closed, and valve-protection caps placed on them? ☐ ☐ ☐
12. Are signs posted reading "DANGER, NO SMOKING, MATCHES, OR OPEN LIGHTS," or the equivalent posted? ☐ ☐ ☐
13. Are cylinders, cylinder valves, couplings, regulators, hoses, and apparatuses kept free of oily or greasy substances? ☐ ☐ ☐
14. Is care taken not to drop or strike cylinders? ☐ ☐ ☐
15. Unless secured on special trucks, are regulators removed and valve-protection caps put in place before moving cylinders? ☐ ☐ ☐
16. Do cylinders without fixed wheels have keys, handles, or non-adjustable wrenches on stem valves when in service? ☐ ☐ ☐
17. Are liquefied gases stored and shipped with valve end up and with valve covers in place? ☐ ☐ ☐

18. Are employees trained never to crack a fuel-gas cylinder valve near sources of ignition? ☐ ☐ ☐
19. Before a regulator is removed, is the valve closed and gas released from the regulator? ☐ ☐ ☐
20. Is red used to identify the acetylene (and other fuel-gas) hose, green for the oxygen hose, and black for inert gas and air hoses? ☐ ☐ ☐
21. Are pressure-reducing regulators used only for the gas and pressures for which they are intended? ☐ ☐ ☐
22. Is open circuit (no-load) voltage of arc welding and cutting machines as low as possible and not in excess of the recommended limits? ☐ ☐ ☐
23. Under wet conditions, are automatic controls for reducing no-load voltage used? ☐ ☐ ☐
24. Is grounding of the machine frame and safety ground connections of portable machines checked periodically? ☐ ☐ ☐
25. Are electrodes removed from the holders when not in use? ☐ ☐ ☐
26. Are employees required to shut off the electric power to the welder when no one is in attendance? ☐ ☐ ☐
27. Is suitable fire extinguishing equipment available for immediate use? ☐ ☐ ☐
28. Is the welder forbidden to coil or loop welding electrode cable around his body? ☐ ☐ ☐
29. Are wet machines thoroughly dried and tested before use? ☐ ☐ ☐
30. Are work and electrode lead cables frequently inspected for wear and damage, and replaced when needed? ☐ ☐ ☐
31. Are cable connectors adequately insulated? ☐ ☐ ☐
32. When the object to be welded cannot be moved and fire hazards cannot be removed, are shields used to confine heat, sparks, and slag? ☐ ☐ ☐
33. Are fire watchers assigned when welding or cutting is performed in locations where a serious fire might develop? ☐ ☐ ☐
34. Are combustible floors kept wet, covered with damp sand, or protected by fire-resistant shields? ☐ ☐ ☐

35. Are personnel protected from possible electric shock when floors are wet? ☐ ☐ ☐
36. Are precautions taken to protect combustibles on the other side of metal walls when welding is underway? ☐ ☐ ☐
37. Before hot work begins, are used drums, barrels, tanks, and other containers thoroughly cleaned and tested such that no substances remain that could explode, ignite, or produce toxic vapors? ☐ ☐ ☐
38. Do eye protection, helmets, hand shields, and goggles meet appropriate standards? ☐ ☐ ☐
39. Are employees exposed to the hazards created by welding, cutting, or brazing operations protected with PPE and clothing? ☐ ☐ ☐
40. Is a check made for adequate ventilation in and where welding or cutting is performed? ☐ ☐ ☐
41. When employees work in confined places, is the atmosphere monitored and environmental tests done, and are means provided for quick removal of welders in case of an emergency? ☐ ☐ ☐

11.3 Compressors and Compressed Air

	YES	NO	N/A
1. Are compressors equipped with pressure-relief valves and pressure gauges?	☐	☐	☐
2. Are compressor air intakes installed and equipped with filters to ensure that only clean, uncontaminated air enters the compressor?	☐	☐	☐
3. Are compressors operated and lubricated according to the manufacturer's recommendations?	☐	☐	☐
4. Are safety devices on compressed air systems checked frequently?	☐	☐	☐
5. Before any repair work is done on the pressure systems of the compressor, is the pressure bled off and the system locked out?	☐	☐	☐
6. Are signs posted to warn of the automatic starting feature of the compressors?	☐	☐	☐
7. Is the belt drive system totally enclosed to provide protection on front, back, top, and sides?	☐	☐	☐

8. Are employees strictly PROHIBITED from directing compressed air toward a person? ☐ ☐ ☐

9. Are employees PROHIBITED from using compressed air at over 29 psi (2.03 kg/cm^2) for cleaning purposes unless they use an approved nozzle with pressure relief and clip guard? ☐ ☐ ☐

10. Are employees PROHIBITED from cleaning clothing with compressed air? ☐ ☐ ☐

11. DO employees use personal protective equipment when using compressed air for cleaning? ☐ ☐ ☐

12. Are high-pressure hoses and connections in good repair? ☐ ☐ ☐

11.4 Gas Cylinders

Compressed Gas   YES   NO   N/A

1. Are cylinders labeled with the identity of the contents and associated hazard warnings? ☐ ☐ ☐

2. Are empty cylinders marked with a tag or sign reading "Empty" or "MT"? ☐ ☐ ☐

3. Are cylinders located or stored in areas where they will not be damaged by passing or falling objects or subject to tampering by unauthorized persons? ☐ ☐ ☐

4. Are cylinders properly stored and transported in a manner (secured upright with a chain or strap located approximately 2/3 high on the cylinder to a wall, hand truck, or bench) to prevent them from creating a hazard by tipping, falling, or rolling? ☐ ☐ ☐

5. Are valves, regulators, gauges, couplings, and hoses compatible with the pressure and contents of the cylinder? ☐ ☐ ☐

6. Are valve protectors always placed on cylinders when the cylinders are not in use? ☐ ☐ ☐

7. Are all valves closed off before a cylinder is moved, when the cylinder is empty, and at the completion of each job? ☐ ☐ ☐

8. Have all employees who may use or change out chlorine cylinders and/or transport, store, or otherwise handle chlorine containers completed an authorized Chlorine Operations Safety course? ☐ ☐ ☐

9. Have all custodial or other personnel who may periodically need to enter workspaces such as laboratories and water-treatment rooms or work in facilities where chlorine is used completed an approved Chlorine Safety Awareness course? ☐ ☐ ☐

Fuel Gas

10. Are valve protectors always placed on cylinders when the cylinders are not in use or connected for use? ☐ ☐ ☐

11. Are all valves closed off before a cylinder is moved, when the cylinder is empty, and at the completion of each job? ☐ ☐ ☐

12. Are low-pressure fuel-gas cylinders checked periodically for corrosion, general distortion, cracks, or any other defect that might indicate weakness or render them unfit for service? ☐ ☐ ☐

13. Does the periodic check of low-pressure fuel-gas cylinders include inspection of the bottom of each cylinder? ☐ ☐ ☐

14. Are cylinders legibly marked to clearly identify the gas contained? ☐ ☐ ☐

15. Are cylinders with water-weight capacity of over 30 lb (13.61 kg) equipped (with means for connecting a valve protector or device, or with a collar or recess) to protect the valve? ☐ ☐ ☐

16. Are compressed-gas cylinders stored in areas that are protected from external heat sources (such as flames, intense radiant heat, electric arcs, or high-temperature lines)? ☐ ☐ ☐

17. Are cylinders located or stored in areas where they will not be damaged by passing or falling objects or be subject to tampering by unauthorized persons? ☐ ☐ ☐

18. Are cylinders stored or transported in a manner to prevent them from creating a hazard by tipping, falling, or rolling? ☐ ☐ ☐

19. Are valve protectors always placed on cylinders when the cylinders are not in use or connected for use? ☐ ☐ ☐

11.5 Machine Guarding and Lockout/Tagout

Machine Guarding

		YES	NO	N/A
1.	Have employees completed required training program to instruct employees on safe methods of machine operation?	☐	☐	☐
2.	Is there adequate supervision to ensure that employees are following safe machine operating procedures?	☐	☐	☐
3.	Is there a regular safety inspection of machinery and equipment?	☐	☐	☐
4.	Is all machinery and equipment kept clean and properly maintained?	☐	☐	☐
5.	Is sufficient clearance provided around and between machines to allow for safe operations, setup and servicing, material handling, and waste removal?	☐	☐	☐
6.	Is equipment and machinery securely placed and anchored to prevent tipping or other movement that could result in personal injury?	☐	☐	☐
7.	Is there a power shutoff switch within reach of the operator's position at each machine?	☐	☐	☐
8.	Can electric power to each machine be locked out for maintenance, repair, or security?	☐	☐	☐
9.	Are the noncurrent-carrying metal parts of electrically operated machines bonded and grounded?	☐	☐	☐
10.	Are foot-operated switches guarded or arranged to prevent accidental actuation by personnel or falling objects?	☐	☐	☐
11.	Are manually operated valves and switches controlling the operation of equipment and machines clearly identified and readily accessible?	☐	☐	☐
12.	Are all emergency stop buttons colored red?	☐	☐	☐
13.	Are all pulleys and belts within 7 ft (2.1336 m) of the floor or working level properly guarded?	☐	☐	☐
14.	Are all moving chains and gears properly guarded?	☐	☐	☐
15.	Are splash guards mounted on machines that use coolant to prevent the coolant from reaching employees?	☐	☐	☐

16. Are methods provided to protect the operator and other employees in the machine area from hazards created at the point of operation, ingoing nip points, rotating parts, flying chips, and sparks? ☐ ☐ ☐

17. Are machine guards secure and arranged so that they do not cause a hazard while in use? ☐ ☐ ☐

18. If special hand tools are used for placing and removing material, do they protect the operator's hands? ☐ ☐ ☐

19. Are revolving drums, barrels, and containers guarded by an enclosure that is interlocked with the drive mechanism so that revolution cannot occur unless the guard enclosure is in place? ☐ ☐ ☐

20. Do arbors and mandrels have firm and secure bearings, and are they free from play? ☐ ☐ ☐

21. Are provisions made to prevent machines from automatically starting when power is restored after a power failure or shutdown? ☐ ☐ ☐

22. Are machines constructed so as to be free from excessive vibration when the largest size tool is mounted and run at full speed? ☐ ☐ ☐

23. If machinery is cleaned with compressed air, is air pressure controlled and PPE or other safeguards utilized to protect operators and other workers from eye and body injury? ☐ ☐ ☐

24. Are fan blades protected with a guard having openings no larger than 0.5 in (1.2700 cm) when operating within 7 ft (2.1336 m) of the floor? ☐ ☐ ☐

25. Are saws used for ripping equipped with anti-kickback devices and spreaders? ☐ ☐ ☐

26. Are radial arm saws arranged such that the cutting head will gently return to the back of the table when released? ☐ ☐ ☐

27. Is all machinery or equipment capable of movement required to be de-energized or disengaged and blocked or locked out during cleaning, servicing, adjusting, or setting up operations? ☐ ☐ ☐

28. If the power disconnect for equipment does not also disconnect the electrical control circuit, are the appropriate electrical enclosures identified and is a means provided to ensure that the control circuit can also be disconnected and locked out? ☐ ☐ ☐

29. Is the locking out of control circuits instead of locking out main power disconnects prohibited? ☐ ☐ ☐

30. Are all equipment control valve handles provided with a means for locking out? ☐ ☐ ☐

31. Does the lockout procedure require that stored energy (mechanical, hydraulic, air, etc.) be released or blocked before equipment is locked out for repairs? ☐ ☐ ☐

32. Are appropriate employees provided with individually keyed personal safety locks? ☐ ☐ ☐

33. Are employees required to keep personal control of their key(s) while they have safety locks in use? ☐ ☐ ☐

34. Is it required that only the employee exposed to the hazard can place or remove the safety lock? ☐ ☐ ☐

35. Is it required that employees check the safety of the lockout by attempting a startup after making sure no one is exposed? ☐ ☐ ☐

36. Are employees instructed to always push the control circuit stop button before reenergizing the main power switch? ☐ ☐ ☐

37. Is there a means provided to identify any or all employees who are working on locked-out equipment by their locks or accompanying tags? ☐ ☐ ☐

38. Are a sufficient number of accident prevention signs or tags and safety padlocks provided for any reasonably foreseeable repair emergency? ☐ ☐ ☐

39. When machine operations, configuration, or size require an operator to leave the control station and part of the machine could move if accidentally activated, is the part required to be separately locked out or blocked? ☐ ☐ ☐

40. If equipment or lines cannot be shut down, locked out, and tagged, is a safe job procedure established and rigidly followed? ☐ ☐ ☐

11.6 Standards (Consensus)

Selected ASME/ANSI B107 Series Standards—Hand Tools

B107.100/6-2002	Combination Wrenches (Inch and Metric Series)
B107.8-2003	Adjustable Wrenches
B107.500/11-2008	Pliers, Diagonal Cutting and End Cutting

B107.12-2004	Nutdrivers
B107.13-2003	Pliers: Long Nose, Long Reach
B107.14-2004	Hand Torque Tools (Mechanical)
B107.600/15-2008	Flat Tip Screwdrivers
B107.500/16-2008	Shears (Metal Cutting, Hand)
B107.500/18-2008	Pliers: Wire Twister
B107.23-2004	Pliers, Multiple Position, Adjustable
B107.600/30-2008	Cross Tip Screwdrivers
B107.33-2002	Socket Wrenches, Impact (Metric Series)
B107.100/39-2002	Open End Wrenches (Inch and Metric Series)
B107.400/41-2008	Nail Hammers
B107.400/42-2008	Hatchets and Axes
B107.410/48-2005	Metal Chisels, Punches, and Drift Pins
B107.410/50-2007	Brick Chisels, Brick Sets, and Star Drills
B107.400/54-2008	Heavy Striking Tools
B107.400/57-2005	Bricklayers Hammers and Prospecting Picks
B107.400/58-2007	Riveting, Scaling, and Tinner's Setting Hammers
B107.100-2008	Wrenches
B107.400-2008	Striking Tools
B107.410-2008	Struck Tools
B107.600-2008	Screwdrivers
B175.1	Power Tools—Gasoline-Powered Chain Saws—Safety Requirements
B175.2	Power Tools—Handheld and Backpack, Gasoline Engine-Powered

Chapter 12

Electrical Systems

12.1 Electrical Safety

Electrical dangers such as shock, electrocution, arc flash, and arc blast will always be present on the job, but proper training and safety strategies can minimize the likelihood of injuries and fatalities.

NFPA 70E—Standard for Electrical Safety in the Workplace—covers the full range of electrical safety issues, from work practices to maintenance, special equipment requirements, and installation.

OSHA 1910 Subpart S and OSHA 1926 Subpart K use the comprehensive information in NFPA 70E to spell out what to do to avoid electrical dangers. NFPA 70E details how to avoid electrical dangers.

	YES	NO	N/A
1. Is work on new and existing energized (hot) electrical circuits PROHIBITED until			
• All power is shut off?	☐	☐	☐
• Grounds are attached?	☐	☐	☐
• An effective lockout/tagout system is in place?	☐	☐	☐
• Frayed, damaged, or worn electrical cords or cables are promptly replaced?	☐	☐	☐
• All extension cords have grounding prongs?	☐	☐	☐
• Flexible cords and cables are protected from damage? (Avoid sharp corners and projections.)	☐	☐	☐
2. Are extension cord sets used with portable electric tools and appliances that ONLY are the three-wire type and designed for hard or extrahard service. Look for some of the following letters imprinted on the casing: S, ST, SO, STO?	☐	☐	☐

3. Are all electric tools and equipment maintained in safe condition and checked regularly for defects and taken out of service if a defect is found? ☐ ☐ ☐
4. Are workers PROHIBITED from bypassing any protective systems or devices, designed to protect employees from contact with electric energy? ☐ ☐ ☐
5. Are all overhead electrical power lines LOCATED and IDENTIFIED? ☐ ☐ ☐
6. Are workers PROHIBITED from using ladders, scaffolds, equipment, or materials within 10 ft (3.05 m) of electrical power lines? ☐ ☐ ☐
7. Are all electric tools properly grounded unless they are of the double-insulated type? ☐ ☐ ☐
8. Are workers PROHIBITED from using non-fuse protected multiple plug adapters? ☐ ☐ ☐
9. Are workers given and required to use the proper protective equipment and tools when working around electrical hazards? ☐ ☐ ☐
10. Is there an effective lockout/tagout procedure for work on electrical circuits and equipment? ☐ ☐ ☐
11. Have workers been advised of the location of hazards and proper protective measures to avoid contact with an energized circuit? ☐ ☐ ☐
12. Are safe work practices (de-energizing live parts, discharging capacitors, lockout, etc.) used to prevent electric shock and other injuries? ☐ ☐ ☐
13. Are portable electric tools and equipment grounded or double-insulated? ☐ ☐ ☐
14. Do electric boxes and fittings have approved covers? ☐ ☐ ☐
15. Are defective, damaged, or frayed electric cords replaced promptly? ☐ ☐ ☐
16. Are ground fault circuit interrupters and/or an assured equipment-grounding program used on construction sites? ☐ ☐ ☐
17. Are electrical installations in hazardous locations approved for those locations? ☐ ☐ ☐
18. Is electrical system regularly checked by someone trained in the National Electrical Code (NEC)? ☐ ☐ ☐

Managing and Preventing Risks

19. Are switchboard markings and updates properly maintained? ☐ ☐ ☐

Note: If they're lacking, voltage may be cut off from the wrong appliance during maintenance. Even changing a fuse is difficult when it has to be done through trial and error. Electricity turned off in an unplanned manner and without advance warning may alone cause damage or pose a danger. Markings are a key factor in risk prevention.

20. Is each electrical panel free of obstruction with a minimum of 36 in (91.54 cm) of access clearance? ☐ ☐ ☐
21. Are wall sockets and dividing boxes intact and according to space classification? ☐ ☐ ☐
22. Are workers PROHIBITED from installing wall sockets where they are likely to be under mechanical strain? ☐ ☐ ☐
23. Are workers instructed to AVOID using extension sockets? ☐ ☐ ☐

Note: An ungrounded plug may not be plugged into a grounded wall socket, nor may a plug be altered to fit the socket.

24. Are extension cords in good condition without splices, deterioration, or damage? ☐ ☐ ☐
25. Are extension cords protected from potential damage and run through walls, ceilings, floors, doorways, windows, or under carpets? ☐ ☐ ☐
26. Are extension cords equipped with a grounding prong? ☐ ☐ ☐
27. Are extension cords used ONLY for temporary purposes? ☐ ☐ ☐
28. Are extension cords removed when they are not needed? ☐ ☐ ☐
29. Are workers PROHIBITED to alter the structure of an extension cord by filing or exchanging a regular plug with a grounded plug? ☐ ☐ ☐

Note: If this is done, it is possible to plug an ungrounded extension cord into a grounded socket. The metal frame or shell of an electric appliance plugged into an extension cord may, when malfunctioning, become energized and cause an immediate, life-threatening danger.

30. Are workers PROHIBITED from attaching high-duty appliances to multiple plug extension cords that may cause overload and result in blowing fuses? ☐ ☐ ☐

Note: Overheated electric appliances, including extension cords, are a sign of malfunction or overloading, which create a danger of fire or accident.

31. Is dust and dirt regularly cleaned from machines and equipment to prevent possible heating up and posing a fire hazard? ☐ ☐ ☐
32. Are broken connection boxes replaced to prevent an appliance from causing possible electric shock? ☐ ☐ ☐
33. Is the condition of electric appliances checked regularly to ensure that the junctions are tight and appropriate and that the junction cord is intact? ☐ ☐ ☐

Note: Make sure, by hand, that they are not too hot, thus preventing the risk of short circuiting.

34. Are all lights intact and clean? ☐ ☐ ☐
35. Are lights placed in a location where they can be repaired and serviced? ☐ ☐ ☐
36. Is the enclosure class of the light appropriate? ☐ ☐ ☐

Note: When collected inside a light, dust and dirt can ignite.

Recommendations

Electricity is an inherently dangerous but versatile source of energy. When it is used according to sound safety principles, personal injury and property damage can be effectively prevented.

1. DO NOT use a bulb that exceeds the recommended level for the lamp.
2. OBEY the safety distance marked on the light.
3. When changing fluorescent lamps, always change fuse starter as well; it poses a fire hazard when worn out.

Note: Blinking fluorescent lights cause the light's connector device to heat up and thus create a fire hazard.

4. MAKE SURE that the cabling is intact and appropriately attached.
5. Regular cleaning not only creates a pleasant workplace but also enhances electrical and industrial safety.

Note: While cleaning, the condition of appliances can easily be observed.

6. CHECK that there is a sufficient amount of initial fire extinguishing equipment.
7. MAKE SURE its location is clearly marked.
8. MAKE SURE the extinguishing equipment is suited for the purpose.
9. Seal and make fireproof immediate openings and holes created during repair work.

Note: Sealing solution for leads should be selected so that sealing can be performed as early as possible.

10. EQUIP electric equipment and cable rooms with smoke detectors, which alarm at the early stages of the fire (as when only smoldering) before significant damage has occurred.
11. EQUIP large or important (or both) electric equipment and cable rooms with automatic fire extinguishing equipment.

Note: Gas extinguishing equipment is the most appropriate type, but water sprinklers are also suitable, especially for cable rooms.

12. MAKE SURE, when contracting electrical work, that the fitter is qualified and authorized to perform electrical work.
13. OBTAIN appropriate documents for the electrical work performed and, if necessary, the minutes of the commissioning inspection.

Notes: Lack of notes, blueprints, or switchboard diagrams causes problems when additions are made at a later point.

14. MAKE SURE target it dead, from all directions, before performing any electrical work on it–and when continuing after a break.
15. VERIFY that voltage is disconnected and reliably prevented from being switched on during work in progress.
16. Switch voltage back on after the work is completed—only after confirming that it is safe to do so.

Note: Live work is allowed under certain, very specific conditions and only a person trained for such tasks may perform it. There should be written instructions for the work, and live working tools and equipment must be used.

17. PROHIBIT workers from overloading electrical circuits any time.
18. MAKE SURE all systems are installed as intended by the manufacturer and in accordance with the National Electrical Code and local electrical codes.
19. CLEARLY MARK each electrical circuit breaker or fuse with the name of the electric appliance served by that circuit.

Note: Breaker or fuse identification allows for interruption of the electric current to the circuit in the event of an emergency due to electric shock or faulty appliances, and assists in identification of circuit overloading. MAKE SURE breaker boxes are accessible at all times.

20. Use extension cords with portable tools that are only of the three-wire type with three-prong plugs—except when using double-insulated tools.
21. MAKE SURE plugs are nonconductive.
22. CHECK that the wire sizes of extension cords are capable of handling the load without heating.
23. PROHIBIT the use of multiple plug-on attachments on extension cords.
24. ENSURE that all extension cords are serviceable and free of exposed wiring and splices, frayed areas, and deteriorated insulations (or both).
25. Use extension cords ONLY for temporary purposes, NOT for permanent installation.
26. Install an electrical outlet, where there is a permanent need for an outlet.

Note: While in use, DO NOT place the extension cord in such a manner as to present a tripping hazard.

27. DO NOT place extension cords under rugs, carpets, or mats.
28. DO NOT place extension cords where they may be damaged by foot or vehicular traffic.
29. DO NOT run extension cords through holes in walls, or ceilings, or through doorways, or windows.

30. MAKE SURE all electrically operated equipment, appliances, and tools are effectively grounded.

Note: Grounding may be done in the following ways:
By an approved grounding-type attachment plug provided, the ground wire is attached to the metal enclosure of the electrical conductors and to the plug connected to the device's cord—the ground wire MUST be attached at all times while the equipment is connected to the power supply.
By means of a grounding conductor run with the power conductors in the flexible cord connected to the device with the ground connection always making contact with the electrical outlet ground while the device is connected to the power supply conductors.

Exception: Equipment, appliances, or tools protected by an approved system of double insulation, or its equivalent, need not be grounded.

31. VERIFY that all electric devices bear the label of a nationally recognized testing laboratory such as the Underwriter's Laboratory (UL) or the Factory Mutual Engineering Corp. (FM).
32. CONDUCT periodic inspections of all equipment, appliances, and tools to ensure that all cords are free of wear and splices, and that the casing or insulating covering is free of cracks, holes, or other damage.
33. ENSURE that all covers are in place and that circuit interlocks are functioning where applicable.
34. PULL from service and submit for repair by a qualified person any electric device that is damaged, malfunctioning, or shows signs of unusual, excessive heating or producing "burning" odors.
35. MAKE SURE electrical wall outlets are free of cracks, breaks, or other obvious damage.

Note: Report immediately any outlet damage to Project Control for repair.

36. REMOVE from service immediately any equipment, appliance, or tool that produces shock, no matter how small—for repair by a qualified electrician before returning to service.
37. AVOID excess bending, stretching, and kinking of electric supply cords.
38. NEVER jerk electric cords from wall outlets.
39. NEVER staple, tack, or nail electric cords to walls or floors; insulation can be damaged by these devices, exposing bare wires.

12.2 Safe Work Practices

Workers

	YES	NO	N/A
1. Are workplace electricians familiar with OSHA electrical safety rules?	☐	☐	☐
2. Is compliance with OSHA rules required on all contract electrical work?	☐	☐	☐
3. Are employees prohibited from working alone on energized lines or equipment over 600 V?	☐	☐	☐
4. Are employees forbidden to work closer than 10 ft (3.05 m) from high-voltage (over 750 V) lines?	☐	☐	☐
5. Are all employees required to report (as soon as practical) any obvious hazard to life or property observed in connection with electric equipment or lines?	☐	☐	☐
6. Are employees instructed to make preliminary inspections and/or appropriate tests to determine what conditions exist before starting working on electric equipment or lines?	☐	☐	☐
7. Are only certified electricians allowed to repair electric equipment?	☐	☐	☐

Equipment and Systems

	YES	NO	N/A
8. Does all electric equipment bear a UL or FM or other appropriate label?	☐	☐	☐
9. Is all electric equipment sound, without visible damage, excessive heating, or burning odor?	☐	☐	☐
10. Is electric equipment free from recognized hazards which are likely to cause death or serious physical harm to employees?	☐	☐	☐
11. Does any electric equipment cause any degree of shock when touched?	☐	☐	☐
12. Is all equipment which may arc, spark, or flame kept separated and isolated from combustible material?	☐	☐	☐
13. Is there sufficient clearance around electric equipment for workers to move about freely?	☐	☐	☐
14. Are all equipment cords free of signs of wear or splices?	☐	☐	☐
15. Are all equipment, appliance, and tool cords and extension cords free of exposed wiring and splices?	☐	☐	☐

16. Are necessary switches opened, locked out, and tagged, when electric equipment or lines are to be serviced, maintained, or adjusted ☐ ☐ ☐
17. Are portable handheld electrical tools and equipment grounded or of the double-insulated type? ☐ ☐ ☐
18. Are electrical appliances such as vacuum cleaners, polishers, and vending machines properly grounded? ☐ ☐ ☐
19. Are any electric cords bent, stretched, or kinked? ☐ ☐ ☐
20. Is the use of extension cords PROHIBITED in lieu of a permanent installation? ☐ ☐ ☐
21. Are the wire sizes of any extension cords capable of handling the load without heating? ☐ ☐ ☐
22. Do any appliance or extension cords present a tripping hazard? ☐ ☐ ☐
23. Do extension cords have a grounding conductor? ☐ ☐ ☐
    - Are multiple plug adaptors prohibited? ☐ ☐ ☐
24. Are all extension cords, used for portable equipment, of the three-wire type with three-prong plugs? ☐ ☐ ☐
25. Are ground-fault circuit interrupters (GFCIs) installed on each temporary 15, 20, or 30 A, 125-V alternating current (AC) circuit at locations where construction, demolition, modifications, alterations, or excavations are being performed? ☐ ☐ ☐
    - Is an assured equipment-grounding conductor program in place? ☐ ☐ ☐
26. Are all temporary circuits protected by suitable disconnecting switches or plug connectors at the junction with permanent wiring? ☐ ☐ ☐
27. Are any electrical circuits overloaded by use of either expansion device or extension cords? ☐ ☐ ☐
28. Is each electrical circuit breaker or fuse clearly marked with the name(s) of the electrical appliance served by that breaker or fuse? ☐ ☐ ☐
29. Are all insulating covers free of cracks, holes, or other signs of damage? ☐ ☐ ☐
30. Are all electrical wall outlets sound and free of cracks, breaks, or other signs of damage? ☐ ☐ ☐
31. Is exposed wiring and cords with frayed or deteriorated insulation repaired or replaced promptly? ☐ ☐ ☐

Inspections

32. Are flexible cords and cables free of splices or taps? ☐ ☐ ☐
33. Are clamps or other securing means provided on flexible cords or cables at plugs, receptacles, tools, equipment, and is the cord jacket securely held in place? ☐ ☐ ☐
34. Are all cords, cable, and raceway connections intact and secure? ☐ ☐ ☐
35. Are electrical tools and equipment appropriate, or otherwise protected, for the use in wet or damp locations? ☐ ☐ ☐
36. Are electrical power lines and cables located (overhead, underground, under floor, other side of walls) before digging, drilling, or similar work begins? ☐ ☐ ☐
37. Is the use of metal measuring tapes, ropes, hand lines, or similar devices with metallic thread woven into the fabric prohibited where these could come into contact with energized parts of equipment or circuit conductors? ☐ ☐ ☐
38. Is the use of metal ladders prohibited in areas where the ladder or the person using the ladder could come into contact with energized parts of equipment, fixtures, or circuit conductors? ☐ ☐ ☐
39. Are all disconnecting switches and circuit breakers labeled to indicate their use or equipment served? ☐ ☐ ☐
40. Are disconnecting means always opened before fuses are replaced? ☐ ☐ ☐
41. Do all interior wiring systems include provisions for grounding metal parts or electrical raceways, equipment, and enclosures? ☐ ☐ ☐
42. Are all fuse-protected power strips plugged directly into a wall outlet (not "daisy-chained" together)? ☐ ☐ ☐
43. Are all electrical raceways and enclosures securely fastened in place? ☐ ☐ ☐
44. Are all energized parts of electrical circuits and equipment guarded against accidental contact by approved cabinets or enclosures? ☐ ☐ ☐
45. Is sufficient access and working space provided and maintained around all electrical equipment to permit ready and safe operations and maintenance? ☐ ☐ ☐

46. Are all unused openings (including conduit knockouts) of electrical enclosures and fittings closed with appropriate covers, plugs, or plates? ☐ ☐ ☐

47. Are electrical enclosures such as switches, receptacles, and junction boxes provided with tight-fitting covers or plates? ☐ ☐ ☐

12.3 Hazardous Energy Control

Lockout / Tagout   YES   NO   N/A

1. Have all employees who may have to service, repair, or maintain plug- and cord-connected equipment been instructed to unplug it and keep the plug under their immediate control? ☐ ☐ ☐

2. Is there a written lockout/tagout program that is followed if exposure to hazardous energy cannot be eliminated by simply unplugging machinery or equipment while it is being serviced or maintained? ☐ ☐ ☐

3. Have all personnel responsible for administering or supervising unit lockout/tagout programs attended the Hazardous Energy Control for Supervisors training course conducted by a qualified supervisor or third-party service provider? ☐ ☐ ☐

4. Have all employees who work around machinery or equipment subject to lockout/tagout attended a Lockout/Tagout for Affected Employees training course conducted by a a qualified supervisor or third-party service provider? ☐ ☐ ☐

5. Has each unit's written lockout/tagout program been audited by a qualified person within the past year? ☐ ☐ ☐

6. Have all electricians, welders, and other employees who are qualified and authorized to install, service, or work on or near exposed parts of electrical installations, wiring, and related components operating at 50 V or more to ground, and employees who directly supervise employees for whom this training is required, completed an Electrical Safety (Advanced) training course conducted by a qualified supervisor or a qualified third-party service provider? ☐ ☐ ☐

7. Have all employees, and their supervisors, who are not qualified to perform electrical work, but whose job tasks may require them to operate electrical equipment in hazardous areas (e.g., damp or wet conditions) or to work in mechanical rooms, utility tunnels, or similar areas completed the Electrical Safety (Awareness-Level) course conducted by a qualified supervisor or third-party service provider? ☐ ☐ ☐

Electrical Work

8. Do you require compliance with OSHA standards for all contract electrical work? ☐ ☐ ☐
9. Are all employees required to report any obvious hazard to life or property in connection with electrical equipment or lines as soon as possible? ☐ ☐ ☐
10. Are employees instructed to make preliminary inspections and/or appropriate tests to determine conditions before starting working on electrical equipment or lines? ☐ ☐ ☐
11. When electrical equipment or lines are to be serviced, maintained, or adjusted, are necessary switches opened, locked out, or tagged, whenever possible? ☐ ☐ ☐
12. Are portable electrical tools and equipment grounded or of the double-insulated type? ☐ ☐ ☐
13. Are electrical appliances such as vacuum cleaners, polishers, vending machines, etc., grounded? ☐ ☐ ☐
14. Do extension cords have a grounding conductor? ☐ ☐ ☐
15. Are multiple plug adaptors prohibited? ☐ ☐ ☐
16. Are ground-fault circuit interrupters installed on each temporary 15 or 20 A, 120-V AC circuit at locations where construction, demolition, modifications, alterations, or excavations are being performed? ☐ ☐ ☐
17. Are all temporary circuits protected by suitable disconnecting switches or plug connectors at the junction with permanent wiring? ☐ ☐ ☐
18. Do you have electrical installations in hazardous dust or vapor areas? ☐ ☐ ☐
    - If so, do they meet the National Electrical Code (NEC) for hazardous locations? ☐ ☐ ☐

Electrical Systems

19. Are exposed wiring and cords with frayed or deteriorated insulation repaired or replaced promptly? ☐ ☐ ☐
20. Are flexible cords and cables free of splices or taps? ☐ ☐ ☐
21. Are clamps or other securing means provided on flexible cords or cables at plugs, receptacles, tools, equipment, etc., and is the cord jacket securely held in place? ☐ ☐ ☐
22. Are all cord, cable, and raceway connections intact and secure? ☐ ☐ ☐
23. In wet or damp locations, are electrical tools and equipment appropriate for the use or location or otherwise protected? ☐ ☐ ☐
24. Is the location of electrical power lines and cables (overhead, underground, under floor, other side of walls, etc.) determined before digging, drilling, or similar work is begun? ☐ ☐ ☐
25. Are metal measuring tapes, ropes, hand lines, or similar devices with metallic thread woven into the fabric prohibited where they could come in contact with energized parts of equipment or circuit conductors? ☐ ☐ ☐
26. Is the use of metal ladders prohibited where the ladder or the person using the ladder could come in contact with energized parts of equipment, fixtures, or circuit conductors? ☐ ☐ ☐
27. Are all disconnecting switches and circuit breakers labeled to indicate their use or equipment served? ☐ ☐ ☐
28. Are disconnecting means always opened before fuses are replaced? ☐ ☐ ☐
29. Do all interior wiring systems include provisions for grounding metal parts of electrical raceways, equipment, and enclosures? ☐ ☐ ☐
30. Are all electrical raceways and enclosures securely fastened in place? ☐ ☐ ☐
31. Are all energized parts of electrical circuits and equipment guarded against accidental contact by approved cabinets or enclosures? ☐ ☐ ☐
32. Is sufficient access and working space provided and maintained around all electrical equipment to permit ready and safe operations and maintenance? ☐ ☐ ☐

33. Are unused openings in electrical enclosures and fittings, such as conduit knockouts, junction boxes, and electrical panels, protected by appropriately installed covers, plugs, or plates? ☐ ☐ ☐

34. Are electrical enclosures and fittings such as switches, receptacles, junction boxes, etc., provided with tight-fitting covers, plugs, or plates? ☐ ☐ ☐

35. Are all exterior receptacles, and interior receptacles in wet/damp locations or within 6 ft (1.83 m) of a sink or other water source, equipped with ground-fault circuit interrupter protection? ☐ ☐ ☐

36. Are disconnecting switches for electrical motors in excess of 2 hp able to open the circuit when the motor is stalled without exploding? ☐ ☐ ☐
    - Are switches horsepower rated equal to or in excess of the motor rating? ☐ ☐ ☐

37. Is low-voltage protection provided in the control device of motor driving machines or equipment that could cause injury from inadvertent starting? ☐ ☐ ☐

38. Is each motor disconnecting switch or circuit breaker located within sight of the motor control device? ☐ ☐ ☐

39. Is each motor located within sight of its controller or is the controller disconnecting means able to be locked open or is a separate disconnecting means installed in the circuit within sight of the motor? ☐ ☐ ☐

40. Is the controller for each motor that exceeds 2 hp rated equal to or above the rating of the motor it serves? ☐ ☐ ☐

41. Are employees who regularly work on or around energized electrical equipment or lines instructed in cardiopulmonary resuscitation (CPR)? ☐ ☐ ☐

42. Are employees prohibited from working alone on energized lines or equipment over 600 V? ☐ ☐ ☐

12.4 Standards (Consensus)

ANSI

    C33.27-74   Safety Standard for Outlet Boxes and Fittings for Use in Hazardous Locations

    C2-81   National Electrical Safety Code (Installations of more than 1000 V.)

(OSHA electrical safety regulation 1910 Subpart S refers to ANSI standards involving electrical safety.)

IEEE
1584, 2002 Guide for Arc Flash Hazard Calculations

NFPA
70E 2009 Standard for Electrical Safety in the Workplace
(OSHA bases its electrical safety mandates—OSHA 1910 Subpart S and OSHA 1926 Subpart K—on this NFPA Standard. The OSHA regulations, however, are LAW, and failure to follow these standards could result in a citation, a work shutdown, and fines or other sanctions.)

NIOSH
Electrical Safety—Safety and Health for Electrical Trades Student Manual provides useful safety information and is available for downloading in portable document format.

OSHA
29 CFR 1910.335 (a)(2)(ii) requires the use of protective shields, protective barriers, or insulating materials to protect employees from shock, burns, or other electrically related injuries while working near exposed energized parts or where dangerous electric heating or arcing might occur.

Chapter 13

Worker Protection

Effective management of worker safety and health protection is a decisive factor in reducing the extent and severity of work-related injuries and illnesses and related costs.

OSHA advises employers and encourages them to institute and maintain in their establishments a program that provides adequate systematic policies, procedures, and practices to protect their employees from, and allow them to recognize, job-related safety and health hazards.

An effective program includes provisions for the systematic identification, evaluation, and prevention or control of general workplace hazards, specific job hazards, and potential hazards that may arise from foreseeable conditions.

Although compliance with the law, including specific OSHA standards, is an important objective, an effective program looks beyond specific requirements of law to address all hazards in an effort to help prevent injuries and illnesses, whether or not compliance is an issue.

The extent to which the program is described in writing is less important than how effective it is in practice. As the size of a worksite or the complexity of a hazardous operation increases, however, the need for written guidance increases to ensure clear communication of policies and priorities as well as a consistent and fair application of rules.

Major elements

An effective occupational safety and health program will include four main elements:

- Management commitment and employee involvement—Provides the motivating force and the resources for organizing and controlling activities within an organization. Employee involvement provides

the means by which workers develop or express (or both) their own commitment to safety and health protection for themselves and for their fellow workers.
- Worksite analysis—Involves a variety of worksite examinations to identify existing hazards and conditions and operations in which changes might occur to create new hazards. Effective management actively analyzes the work and worksite to anticipate and prevent harmful occurrences.
- Hazard prevention and control—Where it is not feasible to eliminate such hazards, they must be controlled to prevent unsafe and unhealthful exposure. Elimination or control must be accomplished in a timely manner once a hazard or potential hazard is recognized. Specifically, as part of the program, employers should establish procedures to correct or control present or potential hazards in a timely manner.
- Safety and health training—An essential component of an effective safety and health program helping to identify the safety and health responsibilities of both management and employees at the site. Training is often most effective when incorporated into other education or performance requirements and job practices. Its complexity depends on the size and complexity of the worksite—as well as the characteristics of the hazards and potential hazards at the site.

13.1 Health and Safety

The main project contractor IS responsible for communicating health and safety requirements to subcontractor(s) that they specifically and directly engage prior to the commencement of the project.

	YES	NO	N/A
1. Is a Health and Safety Contractors Handbook provided to the subcontractor(s)?	☐	☐	☐
2. Is a signature required from the subcontractor(s) manager that they have read and understood the requirements?	☐	☐	☐
3. Is a copy of this sent to the project manager/representative, who is to be kept informed by the main contractor as to which subcontractors shall be on-site?	☐	☐	☐
4. Is the main contractor responsible for ensuring that the subcontractor(s) meets and maintains site safety requirements?	☐	☐	☐

5. Is the main contractor responsible for ensuring that when a subcontractor is engaged, the subcontractor is suitably competent, certified, and qualified to undertake the work that will be conducted? ☐ ☐ ☐

6. Does the main contractor, for large or complex projects, provide to project managers any records of site safety meetings and site inductions? ☐ ☐ ☐

7. Are all parties informed that specific hazard management identification and control is to be undertaken by all parties to ensure that the work zone remains safe for project occupants and general public? ☐ ☐ ☐

8. Does the main contractor, at all times, keep the building owner (or owners) management informed of any proposed work occurring that shall or may affect the property or remaining tenancy? ☐ ☐ ☐

13.2 Medical Services/First Aid

Before work begins on a construction site, the site superintendent MUST ensure the medical availability of medical personnel for advice, consultation, and treatment of medical emergencies, and determine the need to designate emergency first-aid responders.

Any employee, acting as a designated first-aid responder or as a good samaritan by rendering emergency first-aid care, MUST receive appropriate follow-up medical treatment by a licensed healthcare provider, immediately after exposure.

Employees MUST be offered the hepatitis B vaccination series (prior to or at the time of exposure).

If employees decline the hepatitis B vaccination series, they must sign the declination form where it will be placed in their personnel file.

Where a medical facility is near the worksite, the employer MUST ensure the following:

	YES	NO	N/A

1. Has the project manager made provisions, BEFORE the project starts, on how to provide prompt medical attention in the event of a serious injury? ☐ ☐ ☐

2. Are the telephone numbers of doctors, hospitals, clinics, and ambulances posted where they are noticeable to all employees? ☐ ☐ ☐

3. Can emergency personnel respond to a site emergency in 3 to 4 minutes where accidents resulting in severe or life-threatening injuries or illnesses, such as suffocation, severe bleeding, or electrocution, could occur? ☐ ☐ ☐

Note: The general rule of thumb is that if the employees are NOT within 4 minutes response time for emergency medical personnel to get to the site, the company needs to train and designate first-aid responders.

4. Can emergency personnel respond within 15 minutes in other circumstances, i.e., where a life-threatening injury is an unlikely outcome of an accident? ☐ ☐ ☐

5. Has a designated responder been trained to render first aid if the travel or response time for emergency response personnel is greater than previously indicated? ☐ ☐ ☐

6. Do designated responders have a current FIRST-AID and CPR card qualifying them to render first aid? ☐ ☐ ☐

Note: The training must be from the U.S. Bureau of Mines, the American Red Cross, or equivalent and MUST be documented.

7. Have the designated first-aid responders been trained in blood-borne pathogens (such as hepatitis, HIV, malaria, etc.)? ☐ ☐ ☐
   - Are the responders offered and given the hepatitis B vaccination series, prior to or at the time of exposure? ☐ ☐ ☐

8. Are first-aid supplies available on each job site with supplies approved by the consulting physician? ☐ ☐ ☐

9. Are there a sufficient number of kits on hand to ensure immediate use? ☐ ☐ ☐

10. Do the first-aid supplies include disposable gloves, disposable mouthpieces for CPR, and eye protection? ☐ ☐ ☐

11. Does the site superintendent check first-aid supplies—before being sent out on each job and at least weekly on each job—to make sure expended items are replaced? ☐ ☐ ☐

13.3 Personal Protective Equipment and Clothing

Equipment	YES	NO	N/A
1. Are there new procedures which require a hazard assessment for PPE?	☐	☐	☐
2. Are protective goggles or face shields provided and worn where there is any danger of flying particles or corrosive materials?	☐	☐	☐
3. Are approved safety glasses required to be worn at all times in areas where there is a risk of eye injuries such as punctures, abrasions, contusions, or burns?	☐	☐	☐
4. Are employees who need corrective lenses (glasses or contacts) in working environments with harmful exposures required to wear only approved safety glasses, protective goggles, or use other medically approved precautionary procedures?	☐	☐	☐
5. Are protective gloves, aprons, shields, or other means provided and required where employees could be cut or where there is reasonably anticipated exposure to corrosive liquids, chemicals, blood, or other potentially infectious materials?	☐	☐	☐

Note: See 29 CFR 1910.1030(b) for the definition of "other potentially infectious materials."

	YES	NO	N/A
6. Are hard hats provided and worn where danger of falling objects exists?	☐	☐	☐
7. Are hard hats inspected periodically for damage to the shell and suspension system?	☐	☐	☐
8. Is appropriate foot protection required where there is the risk of foot injuries from hot, corrosive, poisonous substances; falling objects; and crushing or penetrating actions?	☐	☐	☐
9. Are approved respirators provided for regular or emergency use where needed?	☐	☐	☐
10. Is all protective equipment maintained in a sanitary condition and ready for use?	☐	☐	☐

11. Are there eyewash facilities and a quick drench shower within the work area where employees are exposed to injurious corrosive materials? ☐ ☐ ☐
12. Is special equipment needed for electrical workers available? ☐ ☐ ☐
13. Are food and beverages consumed on the premises ONLY in areas where there is no exposure to toxic material, blood, or other potentially infectious materials? ☐ ☐ ☐
14. Is protection against the effects of occupational noise exposure provided when sound levels exceed those of the OSHA noise standard? ☐ ☐ ☐

Spills

15. Have all employees been trained in adequate work procedures, use and maintenance of protective clothing, and proper use of equipment when cleaning up spilled toxic or other hazardous materials or liquids? ☐ ☐ ☐
16. Is a spill kit available to clean up spilled toxic or hazardous materials? ☐ ☐ ☐
17. Are adequate work procedures, protective clothing, and equipment provided and used when cleaning up spilled toxic or other hazardous materials or liquids? ☐ ☐ ☐
18. Are there appropriate procedures in place for disposing of or decontaminating PPE contaminated with, or reasonably anticipated to be contaminated with, blood or other potentially infectious materials? ☐ ☐ ☐

PERSONAL EQUIPMENT   YES   NO   N/A

1. Have task assessments been performed and documented by supervisors to identify potential hazards requiring the use of personal protective equipment (PPE) (such as eye and face, head, hand, respiratory, hearing, and fall protection)? ☐ ☐ ☐
    - Does assessment of the hazards that might require PPE include a review of injuries? ☐ ☐ ☐
    - Has the assessment been verified through written certification? ☐ ☐ ☐
    - Does it identify the workplace evaluated? ☐ ☐ ☐

2. Do supervisors select appropriate and properly fitted PPE suitable for protection from hazards or the likelihood of hazards being found and ensure affected employees use it? ☐ ☐ ☐
3. Have both the supervisor and the employees been trained in PPE procedures? ☐ ☐ ☐
   - What PPE is necessary for job tasks? ☐ ☐ ☐
   - When workers need it? ☐ ☐ ☐
   - How to properly wear and adjust it? ☐ ☐ ☐
4. Has the training been verified through written certification? ☐ ☐ ☐
5. Does it identify the workplace evaluated? ☐ ☐ ☐
6. Have all employees, who may need to wear personal protective equipment and those who directly supervise employees for whom this training is required, completed a Personal Protective Equipment (non-laboratory) course? ☐ ☐ ☐
7. Are ANSI-approved safety glasses worn ALWAYS in areas where there is risk of eye injury? ☐ ☐ ☐
8. Are protective goggles or face shields provided and worn when there is any danger of flying material or caustic or corrosive materials? ☐ ☐ ☐
9. Have all employees, who may have to charge lead acid batteries, completed a Battery Charging Safety course conducted by a qualified supervisor or third-party provider? ☐ ☐ ☐
10. Are eyewash facilities and a quick-drench shower within a work area where employees are exposed to caustic or corrosive materials? ☐ ☐ ☐
    - Are all eyewashes and safety showers inspected annually by project safety inspector? ☐ ☐ ☐
11. Is protection against the effects of occupational noise exposure provided when sound levels exceed those of the OSHA noise and hearing conservation standard? ☐ ☐ ☐
12. Are approved respirators provided for regular or emergency use where needed? ☐ ☐ ☐
13. Is there a written respirator program? ☐ ☐ ☐
14. Are the respirators inspected before and after each use? ☐ ☐ ☐

15. Is a written record kept of all inspection dates and findings? ☐ ☐ ☐
16. Are protective gloves, aprons, shields, or other protection provided against cuts, corrosive liquids, and chemicals? ☐ ☐ ☐
17. Where employees are exposed to conditions that could cause foot injury are safety shoes required to be worn? ☐ ☐ ☐
18. Are hard hats provided and worn where danger of falling objects exists? ☐ ☐ ☐
19. Are hard hats inspected periodically for damage to the shell and suspension system? ☐ ☐ ☐
20. Do workers who are exposed to vehicular traffic wear reflective, high-visibility garments? ☐ ☐ ☐
21. Have all employees, who may work as flaggers or directly supervise these employees, completed a Flagging and Traffic Control Safety course conducted by qualified supervisor? ☐ ☐ ☐
22. Have all employees been trained in adequate work procedures, use and maintenance of protective clothing, and proper use of equipment when cleaning up spilled toxic or other hazardous materials or liquids? ☐ ☐ ☐
23. Is a spill kit available to clean up spilled toxic or hazardous materials? ☐ ☐ ☐
24. Where employees are exposed to conditions that could cause foot injury are safety shoes required to be worn? ☐ ☐ ☐
25. Are eyewash facilities and a quick-drench shower within a work area where employees are exposed to caustic or corrosive materials? ☐ ☐ ☐
    - Are all eyewashes and safety showers inspected annually by project safety inspector? ☐ ☐ ☐

HEAD AND FACE   YES   NO   N/A

1. Are hard hats provided, required, and worn where danger of falling objects exists? ☐ ☐ ☐
2. Are hard hats inspected periodically for damage to the shell and suspension system? ☐ ☐ ☐

3. Are protective goggles or face shields provided and worn where there is any danger of flying particles or corrosive materials? ☐ ☐ ☐

4. Are approved safety glasses required to be worn at all times in areas where there is a risk of eye injuries such as punctures, abrasions, contusions, or burns? ☐ ☐ ☐

5. Are employees who wear corrective lenses (glasses or contacts) in workplaces with harmful exposures required to wear only approved safety glasses, protective goggles, or use other medically approved precautionary procedures? ☐ ☐ ☐

VISION AND HEARING   YES   NO   N/A

1. Are employees required to complete a Prescription Safety Eyewear program conducted by a a qualified supervisor or third-party provider? ☐ ☐ ☐

2. Are protective goggles or face shields provided and worn when there is any danger of flying material or caustic or corrosive materials? ☐ ☐ ☐

3. Are ANSI-approved safety glasses worn at all times in areas where there's risk of eye injury? ☐ ☐ ☐

4. Are there areas in the workplace where continuous noise levels exceed 85 dB? ☐ ☐ ☐

5. Is there an ongoing preventive health program to educate employees in safe levels of noise, exposures, effects of noise on their health, and the use of personal protection? ☐ ☐ ☐

6. Have activities that may expose employees to hazardous noise levels (i.e., operations using power tools and heavy equipment) been identified? ☐ ☐ ☐
    - Have affected employees been enrolled in a Hearing Conservation Program conducted by a qualified third-party provider? ☐ ☐ ☐

7. Have work areas where noise levels make voice communication between employees difficult been identified and posted? ☐ ☐ ☐

8. Are noise levels measured with a sound level meter or an octave band analyzer, and are records being kept? ☐ ☐ ☐

9. Has isolating noisy machinery from the rest of the operations been tried? ☐ ☐ ☐
10. Have engineering controls been tried to reduce excessive noise? ☐ ☐ ☐
11. Where engineering controls are not feasible, are administrative controls, such as using worker rotation, used to minimize individual employee exposure to noise? ☐ ☐ ☐
12. Is protection against the effects of occupational noise provided when sound levels exceed those of the OSHA noise and hearing conservation standard? ☐ ☐ ☐
13. Is approved hearing protective equipment (noise-attenuating devices) available to every employee working in noisy areas? ☐ ☐ ☐
14. Are workers properly fitted and instructed in the use of ear protectors? ☐ ☐ ☐
15. Are employees in high-noise areas given periodic audiometric testing to ensure that the hearing protection system is effective? ☐ ☐ ☐

Note: To determine maximum allowable levels for intermittent or impact noise, see OSHA noise and hearing conservation rules.

16. Are noise levels measured using a sound level meter or an octave band analyzer, and are you keeping records of these levels? ☐ ☐ ☐
17. Is there a preventive health program provided to educate employees about safe levels of noise and exposure, effects of noise on their health, and use of personal protection? ☐ ☐ ☐
18. Are employees who are exposed to continuous noise level of above 85 dB given periodic audiometric testing to ensure that their hearing protection system is effective? ☐ ☐ ☐
    - Are they retrained annually? ☐ ☐ ☐
19. Is approved hearing protection equipment (noise-attenuating devices) used by every employee working in areas where noise levels exceed 90 dB? ☐ ☐ ☐
20. Are employees properly fitted and instructed in the proper use and care of hearing protection? ☐ ☐ ☐
21. Have work areas in which noise levels make voice communication difficult been identified and posted? ☐ ☐ ☐

RESPIRATORY   YES   NO   N/A

1. Have activities that may expose employees to hazardous levels of dusts, mists, fumes, and vapors (i.e., application of pesticides, changing out chlorine cylinders) and that require the use of respirators been identified?   ☐   ☐   ☐

2. Is there a written respirator program?   ☐   ☐   ☐

3. Have affected employees enrolled in a Respiratory Protection Program conducted by a qualified, third-party provider?   ☐   ☐   ☐

4. Are approved respirators provided for regular or emergency use where needed?   ☐   ☐   ☐

Note: See 29 CFR 1910.134 for detailed information on respirators or check OSHA web site.

5. Are the respirators inspected before and after each use?   ☐   ☐   ☐

6. Is a written record kept of all inspection dates and findings?   ☐   ☐   ☐

BODY, HANDS, AND FEET   YES   NO   N/A

1. Are protective gloves, aprons, shields, or other means provided and required where employees could be cut or where there is reasonably anticipated exposure to corrosive liquids, chemicals, blood, or other potentially infectious materials?   ☐   ☐   ☐

Note: See the OSHA Blood-Borne Pathogens standard, 29 CFR 1910.1030(b), for the definition of "other potentially infectious materials."

2. Is appropriate foot protection required where there is the risk of foot injuries from hot, corrosive, or poisonous substances; falling objects; and crushing or penetrating actions?   ☐   ☐   ☐

3. Where employees are exposed to conditions that could cause foot injury, are safety shoes required to be worn?   ☐   ☐   ☐

SANITIZING EQUIPMENT AND CLOTHING   YES   NO   N/A

1. Is all protective equipment and clothing maintained in a sanitary condition and ready for use?   ☐   ☐   ☐

2. Is required personal protective clothing or equipment able to be cleaned and disinfected easily? ☐ ☐ ☐
3. Are employees prohibited from interchanging personal protective clothing or equipment, unless it has been properly cleaned? ☐ ☐ ☐
4. Are machines and equipment that process, handle, or apply materials that could injure employees cleaned or decontaminated (or both) before being overhauled or placed in storage? ☐ ☐ ☐
5. Are employees prohibited from smoking or eating in any area where contaminants are present that could be injurious if ingested? ☐ ☐ ☐
6. When employees are required to change from street clothing into protective clothing, is a clean change room with a separate storage facility for street and protective clothing provided? ☐ ☐ ☐
7. Are employees required to shower and wash their hair as soon as possible after a known contact with a carcinogen has occurred? ☐ ☐ ☐
8. When equipment, materials, or other items are taken into or removed from a carcinogen-regulated area, is it done in a manner that will not contaminate nonregulated areas or the external environment? ☐ ☐ ☐

13.4 Resources

- NIOSH publications
  - Qualitative Risk Characterization and Management of Occupational Hazards:
  - NIOSH Pocket Guide to Chemical Hazards (NPG)—CD-Rom (Superseded by 2004-103)
  - Construction—NIOSH Fatal Occupational Injury Cost Fact Sheet
  - Safety and Health for Electrical Trades
  - Ergonomic Guidelines for Manual Material Handling
  - Simple Solutions: Ergonomics for Construction Workers
  - Development of a Mobile Manipulator to Reduce Lifting Accidents
  - Protect Employees with an Exposure Control Plan
  - Injuries and Deaths from Falls during Construction and Maintenance of Telecommunication Towers

- Personal protective equipment and clothing
  - Determination of Sound Exposures (DOSEs): Software Manual and Implementation Guide
  - Inquiring Ears Want to Know: A Fact Sheet about Your Hearing Test
  - NIOSH Hearing Loss Simulator Instruction and Training Guide
  - Noise Exposure and Overhead Power Line (OPL) Safety Hazards at Surface Drilling Sites
  - They're Your Ears: Protect Them
  - Wearing Hearing Protection Properly
- Respiratory protection
  - Health Effects of Occupational Exposure to Respirable Crystalline Silica
  - Improve Drill Dust Collector Capture through Better Shroud and Inlet Configurations
  - NIOSH Respirator Selection Logic 2004
  - Respiratory Protection Recommendations for Airborne Exposures to Crystalline Silica
  - Silicosis in Sandblasters
  - Minimizing Respirable Dust Exposure in Enclosed Cabs by Maintaining Cab Integrity
- Workplace solutions
  - Control of Hazardous Dust When Grinding Concrete
  - Preventing Worker Deaths and Injuries from Contacting Overhead Power Lines with Metal Ladders
  - Reducing Hazardous Dust Exposure When Rock Drilling during Construction
  - Reducing Hazardous Dust in Enclosed Operator Cabs during Construction
  - Water Spray Control of Hazardous Dust When Breaking Concrete with a Jackhammer
- Worksite
  - Transportation, Communications, Electric, Gas, and Sanitary Services

Part

3

Protection

Chapter 14

Protection Systems

Worksite security involves more than theft issues; vandalism and injuries are also security concerns that can impact insurance liability. Accountability should be assigned not only at the project level, but also with the job supervisors, foremen, and individual workers. Everyone MUST understand that they are responsible for working safely.

Visitors to the job site during operating hours can lead to additional hazards or injuries. Supervisors and all employees should be trained to keep an eye out for persons that do not belong on the site. All job site visitors MUST report to the job trailer, accompanied by a company employee at all times. Before entering the job site, the visitor MUST be equipped with protective equipment commensurate to the exposures encountered on the job.

After hours, the site may be a target for adventurous children looking for fun that could end in a tragic accident, or for vandals and thieves. It is most important to protect assets from crimes and potential liabilities associated with after-hour visitors.

To protect the public on or around a construction begins with implementing a safety management plan, identifying the hazards, and planning the best methods to eliminate or control the hazards.

- Project safety management plan
  - Identifying the hazards to the public
  - Evaluating the risk of harm those hazards create
  - Defining construction methods
  - Prescribing physical safeguards to avoid or reduce injury and prevent property damage
  - Informing all levels of management of the degree of risk
  - Ensuring appropriate training is implemented

- Assessing hazards
  - Elimination—Removing the hazard from the workplace
  - Substitution—Substituting or replacing a hazard or hazardous work practice with a less hazardous one
  - Isolation—Isolating or separating the hazard or hazardous work practice from people involved in the work and members of the public
- Implementing control measures
  - Providing overhead protection—Gantry or enclosed scaffold.
  - Engineering—May include modifications to plant or equipment (if the hazard cannot be eliminated, substituted, or isolated), such as providing guarding to machinery or equipment.
  - Administrative—Introduce work practices that limit exposure, such as road/footpath closure or working on weekends to limit public exposure.
- Required degree of public protection—When assessing the risks to the public. It is important to analyze who would be at risk and how, such as age factors, physical impairments, etc. Consider the adjoining
  - Properties—Proximity and type of adjoining properties; for example, a school located nearby means a large number of children who may be intrigued by construction sites or a business district means high pedestrian movement.
  - Structures—Construction work that affects the stability of adjoining structures MUST ensure that the structural integrity of the buildings alongside are maintained. Construction methods used may also create hazards, such as sheet piling may provide a noise and vibration nuisance.
  - Roads—Traffic flow, if disruption is expected including volume of traffic flow and times of day of such flow. Planning of site operations becomes imperative with prudent access scheduling to minimize traffic disruption.
- Protection measures—Potentially dangerous conditions, including changes to surface level, excavations, holes and trenches, site works, footpath alterations, etc., that can create hazards to the public, can easily be eliminated or controlled by
  - Erecting barriers around hazards
  - Posting warning signs and installing caution lighting where necessary
  - Providing a traffic/person controller to redirect traffic/people
  - Providing a temporary bypass for traffic/people
  - Providing additional lighting at night
  - Installing temporary fencing, where the construction process breaks into security-fenced areas, such as electrical switchyards, swimming pools, chemical storage yards, etc

Note: Temporary fencing MUST be provided to maintain security, ensure excavations across driveways and roadways are backfilled before the end of a working day or provide safe access across the excavation, maintain vigilance during work breaks (ensure a controller is in position during breaks), and provide level pedestrian and wheelchair access around the area.

- Potential hazards
  - Falling material and debris—Any risk of an object or material used in construction work falling onto people who are likely to be in the vicinity of, but not on, the construction site means the main contractor at the construction site MUST ensure that overhead protection is erected that will catch, deflect, or hold any weight and amount of material or objects that might reasonably be expected to fall onto it. Measures to alleviate the hazard of falling material and debris include
    - Erection of a gantry (overhead protection)
    - Use of enclosed scaffold with catch platforms
    - Erection of hoarding or barricades
    - Use of chutes for discharge of debris
    - Clear area maintained around perimeter of building
    - Scheduling of work to minimize risk
  - Plant and equipment—Movement to, around, and on construction sites creates hazards that MUST be isolated from the public by
    - Enclosing the entire construction site with fencing
    - Installing warning signs and lights
    - Arranging for a controller to redirect traffic/people
    - Providing a temporary bypass for traffic/people
    - Erecting barriers around work area
    - Using spotters working with the plant
    - Locking the site at night
    - Tagging out all plant and equipment
  - Dust and hot work controls include
    - Water to control dust nuisance
    - Welding screens
    - Containment to prevent the passage of dust, sparks, etc
    - Programming of work to suit environmental conditions, such as site clearing on a windy day
    - Stabilizing the ground with spray-on cellulose fiber
  - Vibration and noise—Methods to control noise and vibration (often more of a perceived hazard to the public than actually causing physical damage) include
    - Using attenuated machinery
    - Providing acoustic barriers
    - Using smaller machinery to lessen vibration

- Site visitors—Preventing public access to the site can alleviate the majority of hazards. Effective measures of preventing public access are
  - Providing total site fencing
  - Providing security personnel to site
  - Monitoring access/egress points to site
  - Briefing all staff working in occupied premises on where they may access
  - Posting warning signage
- Physical types of public protection—Separation of the public from the construction work by isolation. Other methods, such as gatemen or signage, should only be used in addition to physical separation.
  - Barricades—Erected adjacent to roads also require warning lights to alert motorist of the hazard during night or inclement weather.
  - Containment—Substantial and fully sheeted screen minimum of 6 ft (1.8 m) in height should be used where a greater measure of protection is necessary or where construction work is of a more permanent nature—often used on a demolition site to exclude members of the public and to prevent debris from spilling or rebounding out of the site.

Note: A 6-ft (1.8-m) high-link mesh is an acceptable alternative to containing if the only requirement is to secure the site from members of the public. The fence should be erected to all elevations of the site.

- Scaffolding—Providing public protection to ensure that no materials or dust will leave the working platform.

Note: Where further protection is required, the scaffold should have at the first stage a catch platform not less than 6.56 ft (2 m) in width projecting at an angle of 45 degrees from the vertical extending along the outside and returned at both ends.

- Night-lights—Providing adequate illumination at both ends and under the platform.

FENCING, LIGHTING, AND SIGNAGE

	YES	NO	N/A
1. Are signs posted around the perimeter of the job site indicating NO TRESPASSING to act as a deterrent?	☐	☐	☐

Note: For the maximum effectiveness, they should be posted ever 20 to 24 ft (6.1–7.62 m).

2. Is fencing installed around the job site? ☐ ☐ ☐

Note: For some areas, a 6-ft (1.83-m) chain link fence with a locking gate may be appropriate. For others, an orange plastic snow fence may suffice. The extent of protection needed is a judgment call on the part of the job manager.

3. Is a good light source in place to help to keep thieves and vandals from entering a job site at night? ☐ ☐ ☐

Note: The light source may be from a street light or from lighting installed strategically on the worksite.

Note: Construction areas, ramps, runways, corridors, offices, shops, and storage areas MUST be lighted to not less than the minimum illumination intensities listed in the following table while any work is in progress.

Minimum Illumination Intensities in Footcandles

Footcandles	Area of operation
5	General construction area lighting
3	General construction areas, concrete placement, excavation, waste areas, accessways, active storage areas, loading platforms, refueling, and field maintenance areas
5	Indoor warehouses, corridors, hallways, and exitways
5	Tunnels, shafts, and general underground work areas

Note: Exception: Minimum of 10 fc is required at tunnel and shaft heading during drilling, mucking, and scaling. Bureau of Mines–approved cap lights shall be acceptable for use in the tunnel heading.

10	General construction plant and shops (e.g., batch plants, screening plants, mechanical and electrical equipment rooms, carpenters shops, active store rooms, living quarters, locker or dressing rooms, mess halls, indoor toilets, and workrooms)
30	First-aid stations, infirmaries, and offices

Source: Occupational Safety and Health Administration, Washington, DC (www.osha.gov)

4. Has personal contact or written memorandum informed neighborhood families of the job, asking them to PROHIBIT their children from accessing the job site? ☐ ☐ ☐

Note: This can also informally solicit a neighborhood watch to keep unauthorized persons away from the site. Include in the written correspondence contact names and phone numbers for easy reporting. This will help not only to reduce liability exposure, but also to improve the contractor's community goodwill.

5. Have local law-enforcement agencies been contacted to let them know when and where work will be done and how to contact the company if they should spot a problem? ☐ ☐ ☐

6. Has a security service been employed to watch the site after work-hours or to do periodic drive-by inspections? ☐ ☐ ☐

7. Is there a key control system in operation? ☐ ☐ ☐
   - Are workers PROHIBITED from storing keys in the open during active working hours? ☐ ☐ ☐
   - Are they PROHIBITED from leaving open locks hanging on a gate, with the key inserted, when padlocks are moved for the day? ☐ ☐ ☐

8. Are locks of substantial strength, such as laminated steel pin tumbler locks, U-locks, and shackle-protected padlocks used? ☐ ☐ ☐
   - Do the locks also have built-in tamper-resistant devices? ☐ ☐ ☐
   - Do the locks feature wireless alarms? ☐ ☐ ☐

9. Are cameras installed around the construction site? ☐ ☐ ☐

Note: From a proactive standpoint, knowing that a site is monitored by camera is a deterrent. On the reactive side, the camera can help identify thieves or intruders and can aid in recovery and prosecution.

10. Is a stationary IT surveillance camera installed that allows remote viewing of the site from any computer or cell phone through the Internet access 24 hours per day?. ☐ ☐ ☐
    - Does the surveillance camera activate based on motion? ☐ ☐ ☐

11. Are wireless security protection systems used on the site to monitor numerous pieces of equipment or areas simultaneously through a system located in a secure enclosed area? ☐ ☐ ☐

Source: Continental Western Group, Des Moines, IA.

14.1 Fencing

Improper material or installation of construction site fencing can have a dramatic effect on the required safety and security. It is most important to verify that the project materials are in compliance with the contract specifications and that the fence is installed properly.

Procurement or project managers may want to consider a mandatory requirement of their reviewing material certifications and shop drawings prior to start of installing the fence. This will ensure that proper products will be installed, and that specific installation guidelines have been provided.

		YES	NO	N/A
1.	Does the fence layout in both height and length coincide with the current site plan?	☐	☐	☐
2.	Are post locations marked for fence elevations and changes in fence direction?	☐	☐	☐
3.	Are changes in fence height identified?	☐	☐	☐
4.	Have the proper posthole depths and diameters been determined and located?	☐	☐	☐
5.	Has the proper size and quantity of fencing material and posts been delivered to the site?	☐	☐	☐
	▪ Has cracked or damaged fencing material been placed aside prior to installation?	☐	☐	☐
6.	Is the vertical alignment of fence within tolerances?	☐	☐	☐

Note: Vertical alignment should be checked and the panel realigned prior to the installation of additional panels.

7.	Have the fence lines been properly marked?	☐	☐	☐
8.	Do the fence lines extend beyond the property lines?	☐	☐	☐

Note: The fence should be 4 to 6 in (10.16–15.24 cm) inside the lines.

9.	Have all zoning regulations and restrictive covenants been satisfied?	☐	☐	☐
10.	Has a valid permit been acquired that allows the fence to be constructed?	☐	☐	☐
11.	Have all underground utilities been marked?	☐	☐	☐
12.	If utilities are close to any fence posts, what will be done to protect underground lines?	☐	☐	☐

13. What is the diameter of the fence postholes? ☐ ☐ ☐
14. Will the holes be a uniform width from top to bottom? ☐ ☐ ☐

Note: "V" shaped holes that are smaller at the bottom than at the top can heave in cold winter weather.

15. Are the postholes as deep as the contract states they will be? ☐ ☐ ☐
16. Do the fence posts extend down into holes as far as the contract states they will? ☐ ☐ ☐
17. How far apart are the fence posts? ☐ ☐ ☐
    - Do they exceed the distance set forth in the contract or specifications? ☐ ☐ ☐
18. Are the posts spaced equally? ☐ ☐ ☐
19. Are the fence posts the same height? ☐ ☐ ☐

Note: Fences should follow the contour of the ground. They should only be level if the ground is level.

20. Is the chain fabric the type and size that was specified? ☐ ☐ ☐
21. Is the fabric stretched tightly? ☐ ☐ ☐
22. Does the fence look crisp? Is it straight and true? ☐ ☐ ☐
23. Do the gates operate smoothly? ☐ ☐ ☐
24. Has all construction debris and extra soil been removed from the job site? ☐ ☐ ☐
25. Has the fencing contractor provided a notarized affidavit stating that all labor and materials are paid up-to-date? ☐ ☐ ☐

Source: Chain Link Manufacturers Institute (CLMI).

14.2 Lighting

No matter what type of project (construction, renovation, or demolition) it is, proper lighting is crucial and should be a priority for workers, general public safety, and site security.

SAFETY LIGHTING   YES   NO   N/A

1. Are 5 fc of illumination provided throughout the general construction area? ☐ ☐ ☐

2. Are the workshops and storerooms provided with 10 fc of illumination? ☐ ☐ ☐
3. Is the first-aid station illuminated to 30 fc? ☐ ☐ ☐
4. Are light guards provided where there is a possibility of breakage? ☐ ☐ ☐
5. Are the light fixtures raised above the workers' heads? ☐ ☐ ☐
6. Are wire guards grounded to the electrical grounding system? ☐ ☐ ☐
7. Are stairways and floor and wall opening areas well illuminated? ☐ ☐ ☐

Source: Maryland Occupational Safety & Health (MOSH), Baltimore, MD.

8. Do lighting fixtures and system meet construction site requirements according to
   - 29 CFR 1926.405 (a)(2)(ii) (E&F)? ☐ ☐ ☐
   - NEC Article [Temporary Lighting—525.21 (B); 527.4 (F)' 527.4 (H)]? ☐ ☐ ☐
   - UL 1088? ☐ ☐ ☐

SECURITY LIGHTING   YES   NO   N/A

1. Are security lights directed onto the areas to be protected and not out into the eyes of potential observers? ☐ ☐ ☐
2. Are the lights bright white to provide food color differentiation, making people easier to identify and license plates easier to read? ☐ ☐ ☐

Note: Although bright white lights are more expensive to keep lit, the difference in visibility is worth the money. A general rule of thumb is observers should be able to easily identify a face at 30 ft (9.144 m).

3. Are power lines kept at a height of 24 ft (7.32 m) or more to deter thieves from cutting power lines? ☐ ☐ ☐
   - Are electric meter boxes well protected to prevent tampering? ☐ ☐ ☐
   - Is there a generator, or some other independent source of power, for the construction security lighting? ☐ ☐ ☐

4. Is there a regular light maintenance program in place □ □ □
   in which someone checks the security lights and
   cleans or replaces lights that no longer work?
   - Are lights enclosed in wire mesh or other shield □ □ □
     that prevents tampering or shattering by vandals
     or other potential intruders?
5. Have all blind spots on the site been eliminated? □ □ □

Note: A night-time job site drive-by is crucial to accurately gauge the effectiveness of the site lighting.

Source: Pro-Vigil, San Antonio, TX.

14.3 Signage

The hazards of construction sites are many. Debris, high voltage, moving equipment, and overhead material handling are only a few of the damages involved in construction work. Bold signs with a clear message can make sure that unauthorized people keep away and that workers take appropriate safety precautions.

Effective construction site signage promotes safety and creates visibility for a construction firm.

	YES	NO	N/A
1. Are signs placed around the physical perimeter of the site to inform the general public of the potential hazards and warning to keep a good distance from the site?	□	□	□
• Private property?	□	□	□
• Restricted admittance?	□	□	□
• NO TRESPASSING?	□	□	□
• Construction in progress?	□	□	□
2. Do signs WARN trespassers to KEEP OUT, to ensure that the untrained do not encounter the dangers of the construction site?	□	□	□
• Warn of danger, caution, restricted area, etc?	□	□	□
• Warn of trenches, excavation, and pits?	□	□	□
3. Do signs WARN the general public about overhead lifting and hoisting of materials?	□	□	□
4. Do signs WARN potential thieves or vandals			
• Legal prosecution of offenders?	□	□	□

- Offer rewards for capture and conviction of thieves and vandals? ☐ ☐ ☐
5. Are facility signs posted, such as Pedestrian, Electrical Safety, Forklift Safety, Hazards Warning, Water Safety, and Vehicle Safety (Stop and Speed Limits)? ☐ ☐ ☐
6. Are signs ANSI or OSHA (or both) recommended? ☐ ☐ ☐

14.4   Barricades

Safety barricades and lights

Where required for protection of workers, public safety, or as required by state laws, substantial barricades MUST be provided for areas where construction and demolition work, excavation and trenching, is being performed.

Safety barricades, however, MUST NOT be used in lieu of required guard rails on temporary bridges crossing trenches, excavations, or other openings.

- For protection of the visually impaired, safety barricades MUST be joined together with 3-in (7.62-cm) yellow CAUTION tape, accordingly:
  - One strand of tape running continuous from barricade to barricade at 42 in (1.07 cm) above grade or mounting level
  - One strand of tape running continuous from barricade to barricade at a height of from 4 to 12 in (10.16–30.48 cm) above grade or mounting level
- In addition battery-operated warning lights shall be maintained on such barricades, whenever visibility is restricted, and at night.

OSHA regulations

1926.203

Signs, signals, and barricades are important, if not critical, to the safety of the construction workers.

29 CFR 1926.200   Accident Prevention Signs and Tags
29 CFR 1926.201   Signaling
29 CFR 1926.202   Barricades

14.5 Standards (Consensus)

- Fencing

Selected ASTM Standards for Fence Materials and Products (10th Edition) includes 42 ASTM standards that cover:

Chain-link and wire fence fabric
Fence construction and installation
Fences for playgrounds and sports and recreational facilities
Fence fittings and protective coatings
Industrial and commercial horizontal slide gates
Fencing materials used in detention and correctional facilities
Design and fabrication of wood fencing

- Lighting

ANSI/ASHRAE/IESNA—Standard 90.1-2007
American National Standard Institute, Washington, DC (www.ansi.org)
Federal Lighting Guidelines
Federal Energy Administration, Washington, DC (www.eere.energy.gov).
Illumination Engineering Society Handbook
Illuminating Society of North America, New York, NY (www.iesna.org)
Lighting and Power Requirements
American National Standard Institute, Washington, DC (www.ansi.org)

- Signage & Barricades

1926.200, Safety Standards for Signs, Signals, and Barricades (2002)
OSHA, Washington, DC (www.osha.gov)
ANSI Standard D6.1-1971
American National Standards Institute, Washington, DC (www.ansi.org)

ANSI Z35.1-1968	Specifications for Accident Prevention Signs
ANSI Z35.2-1968	Specifications for Accident Prevention Tags
ANSI Z535.2	Environmental and Facility Safety Signs
ANSI Z535.3	Criteria for Safety Symbols
ANSI Z535.5	Safety Tags and Barricade Tapes (for Temporary Hazards)
ANSI AS 1725-2003	Chain Link Fabric Security Fencing and Gates (FOREIGN STANDARD)
ASTM F 567	Standard Practice for Installation of Chain Link Fence
CLMI	Guide for the Selection of Line Post Spacing and Size

Chapter 15

Protective Procedures

15.1 Fire Safety

The general contractor or other designee of the building owner shall be responsible for compliance with industry standards to provide a reasonable degree of safety to life and property on the worksite.

	YES	NO	N/A
1. Are permits obtained before performing outdoor burning—as required by local or state ordinances?	☐	☐	☐
2. Have all chemical storage/dispensing operators, maintenance and operation supervisors, and employees who supervise the storage, use, handling, or dispensing of flammable liquids completed the Fire Safety (Flammable Liquids) course conducted by a qualified person or service provider?	☐	☐	☐
3. Have all departmental supervisors completed the Fire Safety (General) course conducted by a qualified person or service provide?	☐	☐	☐
4. Are evacuation maps showing the location of exits, fire extinguishers, emergency eyewashes, and first-aid kits posted?	☐	☐	☐
5. Are all exits free of obstructions?	☐	☐	☐
6. Are all exits marked by a readily visible "EXIT" sign?	☐	☐	☐
7. Where the location of an exit is not immediately apparent, is access to the exit marked with signs indicating the direction toward the exit?	☐	☐	☐

8. Are all doors and passageways that are not an exit or exit way, but may be mistaken as such, identified by a sign reading "NOT AN EXIT"? ☐ ☐ ☐

9. Are all self-closing doors free to close and latch as required (not blocked in the open position)? ☐ ☐ ☐

10. Are all fire extinguishers fully charged; kept in designated, accessible, and unobstructed Locations; and have employees, and have employees completed a Fire Extinguisher Training course within the past 3 years? ☐ ☐ ☐

11. Do all fire alarm boxes and annunciator panels have unobstructed access? ☐ ☐ ☐

12. Do all fire sprinklers have at least 18 in (45.72 cm) of clearance between the sprinkler head deflector and the top of any stored materials? ☐ ☐ ☐

13. Are flammable liquids stored in approved containers and cabinets? ☐ ☐ ☐

14. Are personal space heaters located away from combustibles and equipped with safety switches that will shut off power to the unit immediately if it is tipped over? ☐ ☐ ☐

FIRE PROTECTION YES NO N/A

National Fire Protection Association Standard #1

1. Is the local fire department well acquainted with the job site or facilities, its location, and specific hazards? ☐ ☐ ☐

2. Is there a written fire prevention plant available, if there are 11 or more employees on site? ☐ ☐ ☐

3. Does the plan describe the type of fire protection equipment and/or systems (if any) that are available for use? ☐ ☐ ☐

4. Are practices and procedures established to control potential fire hazards and ignition sources? ☐ ☐ ☐

5. Are workers aware of the fire hazards of the materials and processes to which they are exposed? ☐ ☐ ☐

6. If the site has a fire alarm system, is it certified as required? ☐ ☐ ☐
7. Is the fire alarm system tested at least annually? ☐ ☐ ☐
8. Is proper clearance of 18 in (45.72 cm) maintained below sprinkler heads? ☐ ☐ ☐
9. Are sprinkler heads protected by metal guards, when exposed to physical damage? ☐ ☐ ☐
10. Are automatic sprinkler system, water control valves, and air and water pressure checked weekly/ periodically as required? ☐ ☐ ☐
11. Is the maintenance of automatic sprinkler systems assigned to responsible persons or to a sprinkler contractor? ☐ ☐ ☐
12. Are portable fire extinguishers
    - Provided in adequate numbers and types? ☐ ☐ ☐
    - Mounted in readily assessable locations? ☐ ☐ ☐
    - Recharged regularly and noted on the inspection tag? ☐ ☐ ☐
    - Inspected annually by a service provider and "quick-checked" monthly by staff with documentation entered on the inspection tag? ☐ ☐ ☐
13. If employees are expected to use fire extinguishers and fire protection procedures, are they adequately trained and periodically instructed in both? ☐ ☐ ☐
14. If employees are not trained to use fire extinguishers, are they trained to immediately evacuate the building? ☐ ☐ ☐
15. If you have interior stand pipes and valves, are they inspected regularly? ☐ ☐ ☐
16. If you have outside private fire hydrants, are they flushed at least once a year and on a routine preventive maintenance schedule? ☐ ☐ ☐
17. Are fire doors and shutters
    - Unobstructed and protected against obstructions, including their counterweights? ☐ ☐ ☐
    - In good operating condition? ☐ ☐ ☐
    - Fusible links in place? ☐ ☐ ☐

FLAMMABLE AND COMBUSTIBLE MATERIALS YES NO N/A

1. Are combustible scrap, debris, and waste materials (oily rags, etc.) stored in covered metal receptacles and promptly removed from the worksite? ☐ ☐ ☐
2. Are proper storage methods practiced to minimize the risk of fire, including spontaneous combustion? ☐ ☐ ☐
3. Are approved containers and tanks used for the storage and handling of flammable and combustible liquids? ☐ ☐ ☐
4. Are all connections on drums and combustible liquid piping (vapor and liquid) tight? ☐ ☐ ☐
5. Are all flammable liquids kept in closed containers when not in use (e.g., parts cleaning tanks, pans, etc.)? ☐ ☐ ☐
6. Are bulk drums of flammable liquids grounded and bonded to containers during dispensing? ☐ ☐ ☐
7. Do storage rooms for flammable and combustible liquids have explosion-proof lights and mechanical or gravity ventilation? ☐ ☐ ☐
8. Is liquefied petroleum gas stored, handled, and used in accordance with safe practices and standards? ☐ ☐ ☐
9. Are "NO SMOKING" signs posted on liquefied petroleum gas tanks and in areas where flammable or combustible materials are used or stored? ☐ ☐ ☐
10. Are "NO SMOKING" rules enforced in areas involving storage and use of flammable materials? ☐ ☐ ☐
11. Are liquefied petroleum storage tanks guarded to prevent damage from vehicles? ☐ ☐ ☐
12. Are all solvent wastes and flammable liquids kept in fire-resistant, covered containers until they are removed from the worksite? ☐ ☐ ☐
13. Is vacuuming used whenever possible rather than blowing or sweeping combustible dust? ☐ ☐ ☐
14. Are firm separators placed between containers of combustibles or flammables that are stacked one upon another to ensure their support and stability? ☐ ☐ ☐
15. Are fuel gas cylinders and oxygen cylinders separated by distance and fire-resistant barriers while in storage? ☐ ☐ ☐

16. Are employees trained in the use of fire extinguishers? ☐ ☐ ☐
17. Are fire extinguishers selected and provided for the types of materials in the areas where they are to be used:
    - Class A—Ordinary combustible material fires? ☐ ☐ ☐
    - Class B—Flammable liquid, gas, or grease fires? ☐ ☐ ☐
    - Class C—Energized-electrical equipment fires? ☐ ☐ ☐
18. Are appropriate fire extinguishers mounted in storage areas containing flammable liquids so that employees do not have to travel more than
    - 75 ft (22.86 m) of outside areas? ☐ ☐ ☐
    - 10 ft (3.048 m) of any inside areas? ☐ ☐ ☐
    - 75 ft for a class A fire storage area? ☐ ☐ ☐
    - 50 ft for a class B fire storage area? ☐ ☐ ☐
19. Are all extinguishers fully charged and in their designated places? ☐ ☐ ☐
20. Are fire extinguishers free from obstructions or blockage? ☐ ☐ ☐
21. Is a record maintained of required monthly checks of extinguishers? ☐ ☐ ☐
22. Are all extinguishers serviced, maintained, and tagged at intervals not to exceed 1 year? ☐ ☐ ☐
23. Where sprinkler systems are permanently installed, are the nozzle heads so directed or arranged that water will not be sprayed into operating electrical switchboards and equipment? ☐ ☐ ☐
24. Are safety cans used for dispensing flammable or combustible liquids at point of use? ☐ ☐ ☐
25. Is the transfer/withdrawal of flammable or combustible liquids performed by trained personnel? ☐ ☐ ☐
26. Are all spills of flammable or combustible liquids cleaned up promptly? ☐ ☐ ☐
27. Are storage tanks adequately vented to prevent the development of excessive vacuum or pressure as a result of filling, emptying, or atmospheric temperature changes? ☐ ☐ ☐

28. Are storage tanks equipped with emergency venting that will relieve excessive internal pressure caused by fire exposure?   ☐ ☐ ☐

29. Are rules enforced in areas involving storage and use of hazardous materials?   ☐ ☐ ☐

15.2 Hazardous and Toxic Materials

OSHA hyperlink "http://www.osha.gov/dts/chemicalsampling/toc/toc_chemsamp.html"—"Chemical Sampling Information (CSI)"; and Environmental Protection Agency's (EPA's) Toxic Substance Control Act (TSCA) Chemical Substances.

HAZARDOUS CHEMICAL EXPOSURE   YES   NO   N/A

1. Are employees aware of the potential hazards and trained in safe handling practices for situations involving various chemicals stored or used in the workplace such as acids, bases, caustics, epoxies, phenols, etc?   ☐ ☐ ☐

2. Is employee exposure to chemicals kept within acceptable levels?   ☐ ☐ ☐

3. Are eyewash fountains and safety showers provided in areas where corrosive chemicals are handled?   ☐ ☐ ☐

4. Are all containers, such as vats, storage tanks, etc., labeled as to their contents, e.g., "CAUSTICS"?   ☐ ☐ ☐

5. Are all employees required to use personal protective clothing and equipment when handling chemicals (gloves, eye protection, respirators, etc.)?   ☐ ☐ ☐

6. Are flammable or toxic chemicals kept in closed containers when not in use?   ☐ ☐ ☐

7. Are chemical piping systems clearly marked as to their content?   ☐ ☐ ☐

8. Where corrosive liquids are frequently handled in open containers or drawn from storage vessels or pipelines, are adequate means readily available for neutralizing or disposing of spills or overflows and performed properly and safely?   ☐ ☐ ☐

9. Are standard operating procedures established and are they being followed when cleaning up chemical spills?   ☐ ☐ ☐

10. Are respirators stored in a convenient, clean, and sanitary location, and are they adequate for emergencies? ☐ ☐ ☐

11. Are employees prohibited from eating in areas where hazardous chemicals are present? ☐ ☐ ☐

12. Is personal protective equipment (PPE) used and maintained whenever necessary? ☐ ☐ ☐

13. Are there written standard operating procedures for the selection and use of respirators where needed? ☐ ☐ ☐

14. If a respirator protection program is available, are all workers instructed on the correct usage and limitations of the respirators? ☐ ☐ ☐

15. Are the respirators National Institute for Occupational Safety and Health (NIOSH) approved for this particular application? ☐ ☐ ☐

16. Are they regularly inspected, cleaned, sanitized, and maintained? ☐ ☐ ☐

17. If hazardous substances are used in project work, is there a medical or biological monitoring system in operation? ☐ ☐ ☐

18. Does management know and understand the threshold limit values or permissible exposure limits of airborne contaminants and physical agents used in the workplace? ☐ ☐ ☐

19. Have appropriate control procedures been instituted for hazardous materials, including safe handling practices and the use of respirators and ventilation systems? ☐ ☐ ☐

20. Whenever possible, are hazardous substances handled in properly designed and exhausted booths or similar locations? ☐ ☐ ☐

21. Is a general dilution or are local exhaust ventilation systems used to control dusts, vapors, gases, fumes, smoke, solvents, or mists that may be generated in the workplace? ☐ ☐ ☐

22. Is operational ventilation equipment provided for removal of contaminants from production grinding, buffing, spray painting, and/or vapor degreasing? ☐ ☐ ☐

342 Protection

23. Do employees complain about dizziness, headaches, nausea, irritation, or other factors of discomfort when they use solvents or other chemicals? ☐ ☐ ☐
24. Is there a dermatitis problem? ☐ ☐ ☐
    - Do employees complain about dryness, irritation, or sensitization of the skin? ☐ ☐ ☐
    - Has consideration been given to having an industrial hygienist or environmental health specialist to evaluate the work operations? ☐ ☐ ☐
25. If internal combustion engines are used, is carbon monoxide exhaust kept within acceptable levels? ☐ ☐ ☐
26. Is vacuuming used rather than blowing or sweeping dust whenever possible for cleanup? ☐ ☐ ☐
27. Are materials that give off toxic, asphyxiant, suffocating, or anesthetic fumes stored in remote or isolated locations when not in use? ☐ ☐ ☐

HAZARDOUS SUBSTANCES COMMUNICATION   YES   NO   N/A

1. Have you compiled a list of hazardous substances used in your workplace? ☐ ☐ ☐
2. Is there a material safety data sheet (MSDS) readily available for each hazardous substances or class of substance used in your workplace? ☐ ☐ ☐
3. Is there a written hazard communication program dealing with, and someone responsible for, MSDS labeling and employee training? ☐ ☐ ☐
4. Is each container for a hazardous substance (i.e., vats, bottles, storage tanks, etc.) labeled with product identity and a hazard warning (communication of the specific health hazards and physical hazards)? ☐ ☐ ☐
5. Do you have an employee training program for hazardous substances that includes
    - An explanation of what an MSDS is and how to obtain and use one? ☐ ☐ ☐
    - MSDS contents for each hazardous substance or class of substances? ☐ ☐ ☐
    - An explanation of "A Right to Know"? ☐ ☐ ☐
    - Where employees can review the employer's written hazard communication program, including how to use the labeling system and MSDS? ☐ ☐ ☐

- Location of physical and health hazards in particular work areas and the specific protective measures to be used? ☐ ☐ ☐
- An explanation of the physical and health hazards of substances in the work area, how to detect their presence, and specific protective measures to be used? ☐ ☐ ☐
- How employees will be informed of hazards of nonroutine tasks and hazards of unlabeled pipes? ☐ ☐ ☐
- An explanation of the use and limitations of methods that will prevent or reduce exposure, including appropriate engineering controls, work practices, and PPE? ☐ ☐ ☐
- Information on the types, proper use, location, removal, handling, decontamination, and disposal of PPE? ☐ ☐ ☐
- An explanation of the basis for selection of PPE? ☐ ☐ ☐
- An explanation of signs, labels, and color coding? ☐ ☐ ☐

6. Are employees trained in:
   - How to recognize tasks that might result in occupational exposure? ☐ ☐ ☐
   - How to use work practice, engineering controls and PPE, and their limitations? ☐ ☐ ☐
   - How to obtain information on the types, selection, proper use, location, removal, handling, decontamination, and disposal of PPE? ☐ ☐ ☐
   - Who to contact and what to do in an emergency? ☐ ☐ ☐
7. Do you inform other employers whose employees share a work area with your employees, where hazardous substances are used? ☐ ☐ ☐

15.3 Theft and Vandalism

ANSI Homeland Security Standards Panel (HSSP)

To prevent fire, theft, damage, or destruction of property and injury to personnel, develop and implement a safety and security plan for construction sites where the employees' work consists of procedures established to meet the required elements of a site-specific safety and security standard for a construction site.

Construction sites pose a unique hazard for agencies and employees working in or around them. Employees must be made aware of activities occurring around them and be made aware of safety precautions at all times to prevent an accident from occurring.

Establish theft-prevention policies and clearly communicate them to all workers, including subcontractors and their workers. Hold all supervisors and workers accountable for their part in a theft-prevention program.

DO NOT wait until an equipment theft occurs before establishing reporting procedures. These should include immediate contact with area police and key personnel within a contractor's organization.

Train supervisors and team leaders in the steps to take when a theft is suspected.

Contact local police for training assistance.

Establish a list of names and telephone numbers of key personnel and area police and provide it to each work crew.

	YES	NO	N/A
1. Are supervisors and workers encouraged to promptly report suspicious activity of persons around the worksite?	☐	☐	☐
2. Is there a system in operation to verify the identity of people who deliver packages, materials, and equipment?	☐	☐	☐
3. Are three forms of identification required: a driver's license, a company identification car with a photograph, and a credit card?	☐	☐	☐

Note: Call the person's company and verify employment—if still in doubt.

	YES	NO	N/A
4. Are equipment theft losses monitored and analyzed to identify patterns and to measure the effectiveness of overall efforts?	☐	☐	☐
5. Are work crews periodically visited to review their theft-prevention practices?	☐	☐	☐
6. Is there an inventor control system established for all equipment and tools?	☐	☐	☐
7. Are equipment serial numbers assigned to work teams or supervisors?	☐	☐	☐
8. Are supervisors and team members accountable for the equipment's safe return?	☐	☐	☐
9. Has a documented checkout/check-in system been implemented for all tools and equipment?	☐	☐	☐

10. Are work crews NEVER oversupplied with equipment, but ONLY assigned the equipment they will need for the day? ☐ ☐ ☐

SITE WALK ABOUT   YES   NO   N/A

1. Is equipment locked and immobilized during nonworking hours?
   - Are anti-theft devices attached to equipment, such as steering wheel locks, kill switches, tire and wheel/axle locks, locked hood side plates, and locking fuel caps? ☐ ☐ ☐
   - Are all operating leers, handles, etc., where practicable, locked in place or placed under securely locked covers or lids? ☐ ☐ ☐
2. Is an alarm system used to lock and protect all major pieces of equipment and trailers containing tools? ☐ ☐ ☐
   - Are padlock shields on storage trailers or converted shipping containers installed to make padlocks more tamper resistant and inaccessible to bolt cutters? ☐ ☐ ☐
3. Are all tools, equipment, and attachments double-stamped with an identification number, one conspicuous and the other hidden? ☐ ☐ ☐
4. Are WARNING signs posted on equipment to indicate that identification and serial numbers are recorded? ☐ ☐ ☐
5. Are reward decals put on all equipment? ☐ ☐ ☐
6. Are tools and equipment painted with bright, easily recognizable colors to identify them from a distance? ☐ ☐ ☐
7. Are aerosol-applied "microtagger" thermostat plastic coatings used that contain coded pigments or metal particles? ☐ ☐ ☐
8. Are logos or other identifying marks stenciled or bead-welded on equipment? ☐ ☐ ☐
9. Is equipment stored off premises overnight, unless it is in a secured area? ☐ ☐ ☐
   - Is leasing or renting space at a secure self-storage facility considered if equipment must be stored overnight? ☐ ☐ ☐

10. Is equipment on premise stored in a locked building equipped with double-cylinder dead bolts and monitored alarm? ☐ ☐ ☐
11. Is equipment stored in converted overseas shipping containers that are padlocked and protected by padlocked shields? ☐ ☐ ☐
12. Has using a security guard service been considered if the construction site is located in a high-crime area? ☐ ☐ ☐
    - Or a closed-circuit television surveillance system? ☐ ☐ ☐
    - Or watchdogs when accompanied by a trained handler? ☐ ☐ ☐
13. Is every equipment storage enclosed with a security fence? ☐ ☐ ☐
14. Is the entire construction site securely fenced? ☐ ☐ ☐
    - Is it installed to nationally recognized standards—ASTM F567-93 for installation of chain link fence? ☐ ☐ ☐
    - Is a clear zone of at least 5 ft (1.524 m) maintained around all fencing? ☐ ☐ ☐
15. Are ONLY designated workers allowed access to equipment storage areas? ☐ ☐ ☐
16. Are equipment storage buildings and surrounding areas kept well lit and free of hiding places, such as shrubbery, trees, or other visual obstructions? ☐ ☐ ☐
17. Is nighttime lighting (essential) elevated to eliminate dark areas? ☐ ☐ ☐
    - Is it visible from adjacent streets? ☐ ☐ ☐
    - Is it positioned, where possible, so that it does not limit the view or creates glare problems to persons who routinely check the area? ☐ ☐ ☐
18. Are tool and store depots permanently staffed on large projects? ☐ ☐ ☐
    - Must workers sign in and out when tools are required? ☐ ☐ ☐
19. Are workers and persons visiting the site required to use an identification pass card? ☐ ☐ ☐

Source: Everest National Insurance Co., Liberty Corner, NJ (www.everestnational.com).

15.4 Worksite and Equipment Security

While it is next to impossible to keep professional thieves from stealing property from a construction site, the layout of a construction job site and its corresponding security plan can make it extremely difficult for thieves and virtually impossible for vandals (amateurs), and often can make the difference between controlling losses and suffering costly thefts.

A job site without guards, fencing, adequate lighting, or controlled exits makes a very easy target. There is no perfect security program, however, because job sites in different locations require different protective measures.

Some contractors ignore job site theft, or decide not to take action against it. They simply add stolen property cost to job costs. These direct costs can be substantial. As much as 3 to 5 percent of the job cost, according to one estimate. But there are also some significant indirect costs of job site theft:

- When stolen tools aren't available, delays inevitably occur and productivity drops.
- Contractors sometimes buy cheap tools to cut theft losses. These tools may be too shoddy to steal; however, they also negatively affect productivity—they work poorly, wear out quickly, and break.
- When theft is rampant at a job site, employees' tools are likely to be stolen, and some union contracts hold employers responsible for theft of employees' tools.
- Employer-tolerated theft hurts employee morale; honest workers do not like working where theft is ignored. Poor morale leads to poor productivity and friction.

Source: American Insurance Services Group, Inc.

CONSTRUCTION SITE	YES	NO	N/A
1. Does your receiving area confine materials to a specific area?	☐	☐	☐
2. Does the site have a security fence at least 8 ft in height, enclosing the entire storage area?	☐	☐	☐
3. Does the site have lighting after daylight hours?	☐	☐	☐
4. Have you considered having a watchman patrol the site?	☐	☐	☐
5. Do you provide locked enclosures for storing small high-value items?	☐	☐	☐

6. Are storage sheds or fenced areas used to properly secure all tools and equipment? ☐ ☐ ☐
7. Have you considered
   - Marking your equipment? ☐ ☐ ☐
   - Painting your equipment? ☐ ☐ ☐
8. Have you implemented a good inventory control, including
   - Permanently marked company tools? ☐ ☐ ☐
   - A "check-in and -out" system? ☐ ☐ ☐
   - A daily a record of when and to whom they are issued, and when they are returned? ☐ ☐ ☐
   - Gang boxes and tool sheds locked, at all times? ☐ ☐ ☐
   - Inexpensive die stamps or etching tools to employees, so they can mark their own identification on personal property? ☐ ☐ ☐

EQUIPMENT   YES   NO   N/A

1. Are cabs on all vehicles locked and ignition keys removed when not in use? ☐ ☐ ☐
2. Do you disable machines with hidden ignition cutout switches? ☐ ☐ ☐
3. Are equipment windows fitted with metal shields, where practical? ☐ ☐ ☐
4. Are oil and gas tank caps locked when equipment is stored? ☐ ☐ ☐
5. Is the number of persons having responsibility of "key control" limited? ☐ ☐ ☐
6. Is a log kept current listing the type of key issued, to whom, on what date, and for what purpose? ☐ ☐ ☐
7. Are all unused keys kept under lock? ☐ ☐ ☐
8. Are extra keys kept to a minimum? ☐ ☐ ☐
9. Do you change your locks periodically? ☐ ☐ ☐
10. Have you considered using extra security locks, such as those having changeable combinations? ☐ ☐ ☐

Note: To prevent unauthorized duplication, "plug" keys with a rivet through the bow as a means of preventing alignment are needed for machined duplication.

15.5 Unsafe Site Conditions

OSHA 29 CFR 1926: Safety and Health Regulations for Construction

	YES	NO	N/A
1. Are employees instructed to inspect their work area and equipment before each shift to identify unsafe conditions?	☐	☐	☐
2. Do all project employees keep alert for unsafe conditions?	☐	☐	☐
3. Is an identified, unsafe condition reported immediately to a supervisor?	☐	☐	☐
4. Are employees instructed to follow project chain of command to make sure that appropriate management personnel are informed of the problem?	☐	☐	☐
5. Does the supervisor evaluate the risk of personal injury, public liability, and damage to property or equipment, and initiate steps for correction of unsafe condition?	☐	☐	☐
6. Are employees instructed to call the project's risk management safety office to report an unsafe condition,			
▪ If a supervisor is not available?	☐	☐	☐
▪ If the problem is not adequately corrected in a timely manner?	☐	☐	☐
7. Does the project's risk management supervisor perform, at a minimum, annual inspections of the facilities and worksites?	☐	☐	☐
8. Do all project employees cooperate fully with risk management loss investigations?	☐	☐	☐

15.6 Standards (Consensus)

ANSI/ASEE A 10.34 (R 2005)—Protection of the Public on or Adjacent to Construction Sites

Appendix A—Public Hazard Control Plan (non-mandatory, advisory)

- Specifies enclosing each equipment storage area with a security fence as an essential element in good site security, with all construction sites securely fenced.
- Specifies that security fencing MUST be in accordance with nationally recognized standards ASTM F567-93. (Good fencing, however, requires constant monitoring and maintenance.)

ISO 28000—Supply Chain Security Management Systems Package

- Establishes a security system that will protect people, goods, infrastructure, equipment, and transportation against security incidents and other potentially devastating situations.
- Specifies the requirements to establish, implement, maintain, improve, and audit a security management system.
- Supported with guidelines on creating a security plan, training program, and implementing the security management system.

Appendix A

Bibliography

Aerial Platforms

Aerial Devices Safety Manual, by Association of Equipment Manufacturers
  Contains a wealth of information and many practical job-proven suggestions for encouraging safety-conscious attitudes and performance. Experts worked together to develop safety recommendations for this manual.

Vehicle Mounted Aerial Platform CD, by Arxcis Inc.
  iCD is a complete training program containing PowerPoint® slides with an instructor manual, a student manual, a final exam, an answer key, certificates, OSHA Standards, accident case studies, and an inspection booklet.

Crane and Rigging

Crane & Rigging Large Desk Reference Size, Bob's Overhead (204 pages), by Bob De Benedictis
  Guide covers overhead cranes, detailed inspection procedures, daily checklist, operating practices, weight tables, sling angle and tensions, rigging hardware and hitches.

IPT's Crane & Rigging Handbook (496 pages), by Ron Garby
  Can be used as a training program for riggers, crane operators, in-house maintenance personnel, engineers, etc., or as a toolbox reference. It addresses wire rope and inspection, rigging, hardware, slings and safe working loads, chain data, mobile crane setup, stability, and much more. Complete with 450 illustrations.

"Crane Load Stability Test Code—SAE J765," in *SAE Recommended Practice Handbook*, by Society of Automotive Engineers, Warrendale, Pennsylvania (www.sae.org).

Crane Safety Manual, by Association of Equipment Manufacturers
  Contains information and many practical job-proven suggestions for encouraging safety-conscious attitudes and performance. Experts from equipment manufacturing companies worked together to develop the safety recommendations in this manual.

Cranes and Derricks, 3rd Edition (576 pages), by Howard Shapiro, P.E., Jay Shapiro, P.E., and Lawrence Shapiro, P.E.
  Comprehensive handbook on the selection, installation, and safe use of cranes and derricks on construction sites. This third edition includes information on load handling situations as well as updated safety and accident avoidance guidelines.

Cranes and Hoists, Article 610, NFPA 70, National Electrical Code, National Fire Protection Association, Quincy, Massachusetts

Cranes, Monorails, and Hoist, NUM-1   Rules for Construction of Bridge or Trolley or Hoist of the Underhung Type

Cranes, Monorails and Underhung, B30.11

Cranes, Overhead and Gantry, B30.17   Top Running Bridge, Single Girder, Underhung Hoist

Cranes Overhead and Gantry, B30.2   Top Running Bridge, Single or Multiple Girder, Top Running Trolley Hoist, American Society of Mechanical Engineers, New York, NY

Overhead Traveling cranes
  70   Specifications for Top Running Bridge and Gantry-Type Multiple Girder Electric overhead traveling cranes
  74   Specifications for Top Running and Under running Single Girder Electric, Utilizing Under running Trolley Hoist
       Crane Manufacturer's Association of America, Charlotte, North Carolina

Mobile Crane, Rigging (26 pages)
  Mike's Safety Tips. 3 booklets
  Each booklet contains over 25 tips per subject; rigging, practices, mobile crane operations, and overhead crane use.

Mobile Crane Manual (419 pages)
  Construction Safety Association of Ontario, Toronto, Ontario, Canada
  156 load chart problems from 17 different cranes. Excellent presentation of main components, terminology, quadrants of operation, pre-lift considerations, setup, and operating procedures. 560+ illustrations.

Mobile Crane Operator Reference Card (20 panels), by Training and Inspection Resource Center
   Folding card addresses the following subjects: crane leverage, pre-op inspection, crane setup, operator checklist, electrical clearance, hand signals, weight consideration, weights of materials, sling capacities, hitch types, boom control, hardware, critical lift, reeving, suspended personnel platforms, and hoist rope capacities.

Mobile Cranes (175 pages), by Crane Institute of America
   Combines the practical use of a graphic and text combination making it easy to find and understand the content. Topics include hoisting personnel, load charts, inspection, working around power lines, operating procedures, and much more.

Overhead and Gantry Cranes
   Rules for construction of Top Running Bridge, Multiple Girder, NOG-1. American Society of Mechanical Engineers, New York, NY

Safety and Health Topics: Crane, Derrick and Hoist Safety—Hazards and Possible Solutions
   OSHA references to aid in identifying crane, derrick, and hoist hazards in the workplace. (www.osha.gov/SLTC)

Steel Erection
   Compliance with provisions of the Steel Erection standard, developed with industry and labor through negotiated rule making, could dramatically avert fatal accidents (1 death per 1600 workers) and nearly 1150 annual lost-workday injuries—OSHA eTools and Expert Advisors: (www.osha.gov/eTools)

Floor Openings

Floor and Wall Openings, Railings, and Toe boards
   ANSI 12.1   Safety Code for Railing
   American National Standards Institute/American Society of Mechanical Engineers, New York, NY

Floor and Wall Openings, Stairs, and Railing Systems
   A1264.1   Safety Requirements for Workplace
   American National Standards Institute, New York, NY

Forklift Safety (Industrial Trucks)

Forklift Safety Handbook, by Crane Institute of America
   Provides operators, managers, and safety professionals with the tools to identify and control lift truck related hazards. Uses graphic illustrations to describe lift truck safety issues. It details training

requirements, physical principles, lifting, moving, and placing loads in various operating environments.

Forklift Safety Pocket Guide, by Business and Legal Reports
Topics include safety, inspection, driving, loading, parking, power sources, and maintenance pertaining to forklifts. Booklet includes a dictionary of terms, a tear out inspection checklist, and safety quiz.

Low Lift and High Lift Trucks
B56.1   Safety Standard for low lift and high lift trucks
Nameplates and Markings, B56.7.5
Forks, B56.7.27
Platforms, B56.7.35

High Lift Industrial Trucks
B56.10   Manually Propelled
American Society of Mechanical Engineers, New York, NY

Industrial Trucks
Internal Combustion Engine-Powered, Standard 558 for Safety
Electric Battery-Powered Industrial Trucks, Standard 583 for Safety
Underwriters Laboratories, New York, NY

Powered Industrial Trucks, 505 Fire Safety Standard for
Including Type Designations, Areas of Use, Conversions, Maintenance, and Operation.
National Fire Protection Association, Quincy, Massachusetts

Hazard Communications

Hazard Communication
Hazard Communication: Foundation of Workplace Chemical Safety Programs.
OSHA website index for resources on hazard communication.
(www.osha.gov/dsg)

Hazard Communication Guidelines for Compliance
OSHA Publication 3111 (2000), 112-kb PDF, 33 pages.
This document aids employers in understanding the Hazard Communication standard and in implementing a hazard communication program.
(www.osha.gov/Publications)

Hazard Communication Standard
OSHA Fact Sheet No. 93-26 (1993), 3 pages.
Highlights protections under OSHA's hazard communication standard.
(www.osha.gov/Publications)

Hoists Safety

Hoisting & Rigging Safety Manual (184 pages) by Construction Safety Association of Ontario
  Designed for field and classroom use. Covers major subjects from CSAO's rigging and mobile crane manuals. Sections deal with hoisting and rigging hazards, fiber and wire ropes, rigging tools and devices, slings and hardware, and an introduction to crane operations.

Manually Lever-Operated Hoists, B30.21
  American Society of Mechanical Engineers, New York, NY

Overhead Hoists (Underhung), B30.16

Personnel and Materials Hoist
  Safety Check Logbook and Waterproof Pocket
  Daily pre-use safety checks for personnel and materials hoists. Includes fault and repair details; compulsory service register.

Inspections

Annual Inspection Checklist, by Crane Institute of America
  Annual, periodic, and monthly crane inspection checklists. These multipage checklists are made of heavy card stock and provide the OSHA/ANSI reference for each item inspected. Included are areas for documenting wire rope inspection and load testing.

Inspection Checklists for OSHA Required Inspection (12 checklists per pad), by Crane Institute of America
  Annual, periodic, and monthly crane inspection checklists provide OSHA/ANSI reference for each item inspected. Included are areas for documenting wire rope inspection and load testing.

Jacks and Air Lifts

Jacks, Industrial Rollers, Air Casters, and Hydraulic Gantries, B30.1-2009
  Construction, operation, inspection, testing, and maintenance of mechanical ratchet jacks, hand- or power-operated mechanical screw jacks, hand- or power-operated hydraulic jacks, air lifting bags, industrial rollers, air casters, and telescopic hydraulic gantry systems.

Materials Handling

Materials Handling and Storage
  OSHA Publication 2236 (Revised 2002), 559, 40 pages. Guide to hazards and safe work practices in handling materials. (www.osha.gov/Publications)

Personal Protective Equipment

Fall Protection in Construction
OSHA Publication 3146 (Revised 1998), 43 pages.
(www.osha.gov/Publications)

Personal Protective Equipment
OSHA Publication 3155 (2003), 44 pages. Equipment most commonly used for protection for the head, including eyes and face and the torso, arms, hands, and feet. Includes the use of equipment to protect against life-threatening hazards.
(www.osha.gov/Publications)

Personal Protective Equipment, Safety & Health Topics
OSHA index to hazard recognition, control, and training related to personal protective equipment. (www.osha.gov/SLTC)

Rigging and Lifting

American Rigging & Lifting Handbook
Designed to help reduce industrial accidents in the workplace. Includes best industry practices. Updated guidance on safe operation of equipment with an expanded section on safe operation of cranes. (190 pages)—North Sea Lifting Ltd.

Bob's Overhead Crane & Rigging Handbook for Industrial Operations, 6th Edition (278 pages), by Bob De Benedictis
Large-Desk Reference. Covers the basics on wire rope, slings, fittings, rigging, hitches, and OSHA and ASME/ANSI regulations and codes; includes daily checklist, operating practices, weight tables, sling angle and tensions, rigging hardware, and hitches; focuses on overhead and gantry cranes.

Bob's Rigging & Crane Handbook, 6th Edition (273 pages), by Bob De Benedictis
Contains information on wire rope, rigging equipment, applications and calculations, and mobile crane operational procedures and checklists.

Complete Rigger's Reference Handbook (172 pages), by Mike Riggs
Informational tips on weight estimation, rigging methods, signals and voice commands, rigging capacities, knots, block loading, rigging inspection, and much more in this pocket-sized book.

Handbook for Riggers (128 pages), by W. G. Newberry
A quick reference riggers' guide; includes knots, splices, sling capacities, reeving systems, and good versus poor rigging practices.

Handbook of Rigging 5th Edition (759 pages), by J. A. MacDonald, W. E. Rossnagel, and L. R. Higgins
Now you can own the ultimate guide to designing and operating safe, efficient rigging systems. This brand new edition is a hands-on reference on the subject of lifting, hoisting, and scaffolding for construction and industrial operators.

Handyman "In-Your-Pocket" (768 pages), by Abbeon Cal
Includes information on rope, chain, wire rope, welding, hardware, tools, electrical, construction, weights of materials, and more.

Safety Tips Sheets

No. 1—Proper Use of Hand Signals for Cab-Controlled Cranes
Provides information for employees regarding nine industry standard hand signals that are used for communication between the operator in the crane's cab and the floor person. (2006, September)

No. 2—Preoperational Equipment Check of Cranes and Hoists
Provides information for employees regarding a preoperational equipment checklist for cranes and hoists. (2006, September)

No. 3—Safe Lifting Practices
Provides information for employees regarding safe lifting practices for moving loads of material. (2006, September)

No. 4—Hoist Operation
Provides information for employees on what an operator should and should not do while operating a hoist. (2006, September)

No. 5—Employee Guidelines for Safe Carrying and Transporting Loads
Provides information for employees on guidelines for safe carrying and transporting loads. (2008, May)

No. 6—Management Guidelines for Safe Carrying and Transporting of Loads
Provides information for employers on guidelines for safe carrying and transporting loads. (2008, May)

No. 7—Hazard Evaluation Checklist for Lifting, Carrying, Pushing, or Pulling
Provides employees with a checklist to help identify hazards when moving loads of material. (2008, May)

No. 8—Alternatives to Manual Handling of Individual Containers
Provides information for employees regarding alternatives to manual handling of containers. (2008, May)

No. 9—When and Why to Use Overhead Lifting as Opposed to Manual Lifting
   Provides employees with information on when to use overhead lifting opposed to manual lifting techniques. (2009, December)

No. 10—4 Ways Cranes and Monorails Can Improve Ergonomics
   Provides employees with information on ways cranes and monorails can improve ergonomics. (2009, December)

Source: Occupational Safety and Health Administration

Appendix B

Resources

Government Agencies

CDC   Centers for Disease Control and Prevention
1600 Clifton Road
Atlanta, GA 30333
(800) 232-4636
(www.cdc.gov)

DOE   U.S. Department of Energy
1000 Independence Ave. SW
Washington, DC 20585
(202) 586-5000
(www.energy.gov)

ELCOSH   Electronic Library of
Occupational Health and Safety
Construction Occupational Safety and
Health Information
(www.elcosh.org)

EPA   U.S. Environmental Protection Agency
Ariel Rios Building
1200 Pennsylvania Ave. NW
Washington, DC 20004
(202) 272-0167
(www.epa.gov)

MSHA   Mine Safety and Health Administration
1100 Wilson Boulevard, 21st Floor
Arlington, VA 22209-3939
(202) 693-9400
(www.MSHA.gov)

NIOSH National Institute for Occupational
Safety and Health
395 E Street SW, Suite 9200
Patriots Plaza Building
Washington, DC 20201
(202) 245-0625
(www.cdc.gov)

OSHA Occupational Safety and Health Administration
200 Constitution Ave. NW
Washington, DC 20210
(202) 693-2000
(www.osha.gov)

Standards Developers

ANSI American National Standards Institute
1819 L Street NW, 6th floor
Washington, DC 20036
(202) 293-8020
(www.ansi.org)

American National Standards Institute
25 West 43rd Street, 4th floor
New York, NY 10036
(212) 642-4900
(www.ansi.org)

ASME American Society of Mechanical Engineers
3 Park Avenue
New York, NY 10016-5990
(800) 843-2763
(www.asme.org)

ASSE American Society of Safety Engineers
1800 E. Oakton St.
Des Plaines, IL 60018
(847) 699-2929
(www.asse.org)

ASTM ASTM International
100 Barr Harbor Drive
PO Box C700
West Conshohocken, PA 19428-2959
(610) 832-9500
(www.astm.org)

IEEE	Institute of Electrical and Electronic Engineers 3 Park Avenue, 17th Floor New York, NY 10016-5997 (212) 419-7900 (www.ieee.org)
FM	Factory Mutual 270 Central Avenue PO Box 7500 Johnston, RI 02919-4949 (401) 275-3000 (www.fmglobal.com)
NFPA	National Fire Protection Association 1 Batterymarch Park Quincy, MA 02169-7471 (617) 984-7476 (www.nfpa.org)
SAE	Society of Automotive Engineers 400 Commonwealth Drive Warrendale, PA 15096-0001 (724) 776-4841 (www.sae.org)
UL	Underwriters Laboratories 2600 N.W. Lake Road Camas, WA 98607-8542 (877) 854-3577 (www.ul.com)

Industry Organizations

ACCSH	Advisory Committee on Construction Safety and Health U.S. Department of Labor/OSHA The Curtis Center, Suite 740 West 170 S. Independence Mall West Philadelphia, PA 19106-3309 (215) 861-4000 (www.osha.gov)
ACRP	Association of Crane and Rigging Professionals 28175 Haggerty Road Novi, MI 48377 (800) 690-3921 (www.acrp.net)

AEM	Association of Equipment Manufacturers 6737 W. Washington Street, Suite 2400 Milwaukee, WI 53214-5647 (414) 272-0943 (www.aem.org)
AHSI	American Safety and Health Institute 1450 Westec Drive Eugene, OR 97402 (800) 800-7099 (www.ashinstitute.org)
ALI	American Ladder Institute 401 North Michigan Avenue Chicago, IL 60611 (312) 644-6610 (www.laddersafety.org)
AWRF	Associated Wire Rope Fabricators PO Box 748 Walled Lake, MI 48390 (248) 994-7753 (www.awrf.org)
AWS	American Welding Society 550 N.W. LeJeune Road Miami, FL 33126 (800) 443-9353 (www.aws.org)
BCTD	The Building and Construction Trades Department (AFL-CIO) 815 16th Street, Suite 600 Washington, DC 20006 (202) 347-1461 (www.bctd.org)
CCAA	Crane Certification Association of America PO Box 87907 Vancouver, WA 98687-7907 (800) 447-3402 (www.ccaaweb.net)
CI	Cordage Institute 994 Old Eagle School Road, Suite 1019 Wayne, PA 19087 (610) 971-4854 (www.ropecord.com)

CIAC	Crane Institute of America Certification Inc. 3880 St. Johns Parkway Sanford, FL 32771 (800) 832-2726 (www.CraneInstituteCertification.com)
CICB	Crane Inspection and Certification Bureau PO Box 621388 Orlando, FL 32862 (407) 277-0884 (www.cicb.com)
CLMI	Chain Link Manufacturers Institute 10015 Old Columbia Road #B-216 Columbia, MD 21046 (301) 596-2583 (www.chainlinkinfo.org)
CMAA	Crane Manufacturers Association of America 8720 Red Oak Blvd., Suite 201 Charlotte, NC 28217-3992 (704) 676-1190 (www.mhai.org)
CPWR	The Center for Construction Research and Training (BCTD/AFL-CIO) 8484 Georgia Avenue, Suite 1000 Silver Spring, MD 20910 (301) 578-8500 (www.cpwr.com)
CSAO	Construction Safety Association of Ontario 21 Voyager Court South Etobicoke, ON M9W 5M7 (416) 674-2726 (www.csao.org)
CSC	Construction Safety Council (Mobile Crane Hazard Awareness Guide) 4100 Madison Street Hillside, IL 60162 (708) 544-2082 (www.buildsafe.org)
HIA	Helicopter Association International 1635 Prince Street Alexandria, VA 22314-2818 (703) 683-4646 (www.rotor.com)

HMI	Hoist Manufacturers Institute 8720 Red Oak Blvd., Suite 201 Charlotte, NC 28217-3992 (704) 676-1190 (www.mhia.org)
HTI	Hand Tools Institute 25 North Broadway Tarrytown, NY 10591 (914) 332-0040 (www.hti.org)
HTMA	Hydraulic Tool Manufacturers Association 198 N. Brandon Drive Glendale Heights, IL 60139-2025 (414) 633-3454 (www.htma.net)
IESNA	International Engineering Society of North America 120 Wall Street, Floor 17 New York, NY 10005-4001 (212) 248-5000 (www.ies.org)
ISEA	International Safety Equipment Association 1901 North Moore Street, Suite 808 Arlington, VA 22209-1762 (703) 525-1695 (www.safetyequipment.org)
ITA	Industrial Truck Association 1750 K Street NW, Suite 460 Washington, DC 20006 (202) 296-9880 (www.indtruck.org)
LEEA	Lifting Equipment Engineers Association 3 Osprey Court, Kingfisher Way Hinchingbrooke Business Park Huntingdon, Cambridgeshire PE29 6FN, UK 44 (0) 1480-432801 (www.leea.co.uk)
MHI	Materials Handling Industry of America 8720 Red Oak Blvd., Suite 201 Charlotte, NC 28217-3992 (704) 676-1190 (www.mhia.org)

NACS	National Academy of Construction Safety PO Box 1150 Southampton, PA 18966 (888) 915-7800 (www.nacsgroup.com)
NATE	National Association of Tower Erectors (Checklist for Evaluating Qualified Tower Erection Contractors) 8 Second Street SE Watertown, SD 57201-3624 (www.natehome.com)
NASP	National Association of Safety Professionals 501 Forest Hill Drive PO Box 167 Shelby, NC 28151 (800) 922-2219 (www.naspweb.com)
NCCCO	National Commission for the Certification of Crane Operators 2750 Prosperity Avenue, Suite 505 Fairfax, VA 22031 (703) 560-2391 (www.nccco.org)
NEMA	National Electrical Manufacturers Association 1300 North 17th Street, Suite 1847 Rosslyn, VA 22209 (703) 841-3200 (www.nema.org)
NSC	National Safety Council 1121 Spring Lake Drive Itasca, IL 60143-3201 (630) 285-1121 (www.nsc.org)
NTEA	National Truck Equipment Association 37400 Hills Tech Drive Farmington Hills, MI 48331-3414 (248) 489-7090 (www.ntea.com)
PCSA	Power Crane and Shovel Association 6737 W. Washington Street, Suite 2400 Milwaukee, WI 53214-5647 (414) 272-0943 (www.aem.org)

PTI	Power Tool Institute 1300 Sumner Avenue Cleveland, OH 44115-2851 (www.powertoolinstitute.com)
SCRA	Specialized Carriers and Rigging Association 2750 Prosperity Avenue, Suite 620 Fairfax, VA 22031-4312 (703) 698-0291 (www.scranet.org)
SEAA	Steel Erectors Association of America 2216 W. Meadowview Rd., Ste. 115 Greensboro, NC 27407 (336) 294.8880 (www.seaa.net)
SIA	Scaffold Industry Association 400 Admiral Blvd. Kansas City, MO 64106 (816) 595-4860 (www.scaffold.org)
SSFI	Scaffolding, Shoring, and Forming Institute 1300 Sumner Avenue Cleveland, OH 44115 (216) 241-7333 (www.ssfi.org)
WRC	Welding Research Council PO Box 1942 New York, NY 10136 (216) 658-3847 (www.forengineers.org)
WRTB	Wire Rope Technical Board 801 North Fairfax Street, Suite 211 Alexandria, VA 22314-1757 (703) 299-8550 (www.wrtb.org)
WSTDA	Web Sling and Tie Down Association 2105 Laurel Bush Road #200 Bel Air, MD 21013 (443) 640-1070 (www.wstda.com)

Appendix C

Regulations and Standards

OSHA Regulations

29 CFR 1910—General industry

1910 Subpart D	Walking-Working Surfaces
1910 Subpart F	Powered Platforms, Manlifts, and Vehicle-Mounted Work Platforms
.66 App C	Personal Fall Arrest System (Section I—Mandatory; Sections II and III—Non-mandatory)
.67	Vehicle-mounted elevating and rotating work platforms
.68	Manlifts
1910 Subpart G	Occupational Health and Environment Control
.94	Ventilation
.95	Occupational noise exposure
1910 Subpart I	Personal Protective Equipment
.132	General requirements
.133	Eye and face protection
.134	Respiratory protection
.135	Head protection
.136	Occupational foot protection
137	Electrical protective devices
.138	Hand protection

1910 Subpart N	Materials Handling and Storage
.178	Powered industrial trucks
.179	Overhead and gantry cranes
.180	Crawler locomotive and truck cranes
.181	Derricks
.183	Helicopters
.184	Slings
1910 Subpart P	Hand and Portable Powered Tools and Other Handheld Equipment
1910 Subpart Q	Welding, Cutting, and Brazing
1910 Subpart S	Electrical

29 CFR 1926—Safety and health regulations for construction

29 CFR 1926

Subpart C	General Safety and Health Provisions
.20	General safety and health
.21	Safety training and education
.23	First-aid and medical attention
.24	Fire protection and prevention
.25	Housekeeping
.26	Illumination
.27	Sanitation
.28	Personal protective equipment
.34	Means of egress
Subpart D	Occupational Health and Environmental Controls
.50	Medical services and first aid
.51	Sanitation
.52	Occupational noise exposure
.55	Gases, vapors, fumes, dusts, and mists
.56	Illumination
.57	Ventilation
.59	Hazard communication
.62	Lead
.64	Process safety management of highly hazardous chemicals
.65	Hazardous waste operations and emergency response

Subpart E	Personal Protective and Lifesaving Equipment
.95	Criteria for personal protective equipment
.96	Occupational foot protection
.100	Head protection
.101	Hearing protection
.102	Eye and face protection
.103	Respiratory protection
.104	Safety belts, lifelines, and lanyards
.105	Safety nets
.106	Working over or near water
Subpart F	Fire Protection and Prevention
.150	Fire protection
.151	Fire prevention
.152	Flammable and combustible liquids
.153	Liquefied petroleum gas (LPG).
.154	Temporary heating devices
.156	Fixed extinguishing systems, general
.157	Fixed extinguishing systems, gaseous agent
.158	Fire detection systems
.159	Employee alarm systems
Subpart G	Signs, Signals, and Barricades
.200	Accident prevention signs and tags
.201	Signaling
.202	Barricades
Subpart H	Materials Handling, Storage, Use, and Disposal
.250	General requirements for storage
.251	Rigging equipment for material handling
.252	Disposal of waste materials
Subpart I	Tools—Hand and Power
.301	Hand tools
.302	Power-operated hand tools
.303	Abrasive wheels and tools
.304	Woodworking tools
.305	Jacks—lever and ratchet, screw and hydraulic
.306	Air receivers
.307	Mechanical power-transmission apparatus

Subpart J	Welding and Cutting
.350	Gas welding and cutting
.351	Arc welding and cutting
.352	Fire prevention
.353	Ventilation and protection in welding, cutting, and heating
.354	Welding, cutting, and heating in way of preservative coatings
Subpart K	Electrical

Installation safety requirements

.402	Applicability
.403	General requirements
.404	Wiring design and protection
.405	Wiring methods, components, and equipment, general use
.406	Specific purpose equipment and installations
.407	Hazardous (classified) locations
.408	Special systems

Safety-related work practices

.416	General requirements
.417	Lockout and tagging of circuits

Safety-related maintenance and environmental considerations

.431	Maintenance of equipment
.432	Environmental deterioration of equipment

Consensus Standards

Safety and health program

ASSE/SAFE CONSTRUCTION	Construction Safety Management and Engineering—Comprehensive Safety Resource-Covering Program Essentials, Best Practices, Legal and Regulatory Requirements and Real-World Guidance on Technical Issues
A10.33 (R2004)	Safety and Health Program Requirements for Multiemployer Projects Guidelines for the basic duties of senior contractors and project supervisors in providing a safe construction workplace

A10.38	Basic Elements of an Employer's Program to Provide a Safe and Healthful Work Environment
A10.39	Construction Safety and Health Audit Program
Z490.1	Accepted Practices in Safety, Health, and Environmental Training
Z590.2	Criteria for Establishing Scope and Functions of Professional Safety Position

Rigging worksites

ASSE/SAFE A10.34	Protection of the Public on or Adjacent to Construction Sites
A10.42	Rigging Qualifications and Responsibilities—Safety Requirements
A10.6	Demolition Operations—Safety Requirements

Confined spaces

ASSE/SAFE Z117.1	Safety Requirements for Confined Spaces

Energy lockout/tagout

ASSE/SAFE Z244.1	Control of Hazardous Energy Lockout/Tagout and Alternative Methods
A10.44	Control of Energy Sources (Lockout/Tagout) for Construction and Demolitions Operations

Fire protection

NFPA 70	National Electrical Code

Floor and wall openings

ANSI A10.18	Safety Requirements for Temporary Roof and Floor Holes, Wall Openings, Stairways, and Other Unprotected Edges in Construction and Demolition Operations American National Standard for Construction and Demolition Operations, Railings, and Toe Boards
ANSI A1264.1	Safety Requirements for Workplace Floor and Wall Openings, Stairs, and Railing Systems
ASSE/SAFE A10.24	Roofing—Safety Requirements for Low-Sloped Roofs

NFPA standards

Note: These are NOT OSHA regulations; however, they provide guidance from their originating organizations related to worker protection.

NFPA 30 — Flammable and Combustible Liquids Code (2008)
  Identifies how to properly use, contain, and store flammable and combustible liquids.

NFPA 58 — Liquefied Petroleum Gas Code (2008)
  Identifies requirements for all large tank installations, operating and maintenance procedures, and fire safety analyses.

NFPA 505 — Fire Safety Standard for Powered Industrial Trucks Includes Type Designations, Areas of Use, Conversions, Maintenance, and Operations (2006)
  Identifies industrial truck types for use in hazardous (classified) locations, truck conversions, and maintenance and operation requirements for industrial trucks powered by electric motors or internal combustion engines.

Fire Protection Guide to Hazardous Materials (2001)
  Contains much of the data contained in NFPA documents derived from hundreds of reference sources.

NFPA 49 — Hazardous Chemicals Data
  Identifies 325 chemicals in Material Safety Data Sheet (MSDS) format.

NFPA 325 — Fire Hazard Properties of Flammable Liquids, Gases, and Volatile Solids
  Identifies over 1300 chemicals in tabular format listing various data.

NFPA 491 — Guide for Hazardous Chemical Reactions
  Identifies 3550 dangerous mixtures documented from real-life incidents.

NFPA 497 — Classification of Flammable Liquids, Gases, or Vapors and of Hazardous (Classified) Locations for Electrical Installations in Chemical Process Areas (2008)
  Contains detailed guidelines and diagrams that assist in class I hazardous (classified) area classification for the purpose of properly selecting and installing electric equipment that will not be an ignition source in environments where flammable or combustible liquids, gases, or vapors are processed or handled.

NFPA 704 — Standard for the Identification of the Fire Hazards of Materials for Emergency Response (2007)

Provides a readily recognized and easily understood system, the "Diamond Hazard," for identifying specific hazards and their severity. Hazards are identified using spatial, visual, and numerical methods to describe in simple terms the relative hazards of a material. It addresses the health, flammability, instability, and related hazards that may be presented as short-term, acute exposures that are most likely to occur as a result of fire, spill, or similar emergency.

NFPA 77 Recommended Practice on Static Electricity (2007)
Identifies combustibility parameters and static electric characteristics.

NFPA 430 Code for the Storage of Liquid and Solid Oxidizers (2004)
Identifies oxidizer classifications for 90 chemicals.

NFPA 499 Classification of Combustible Dusts and of Hazardous (Classified) Locations for Electrical Installations in Chemical Process Areas (2008)
Identifies parameters to determine the degree and extent of hazardous locations for dusts, including National Electrical Code (NEC) groups.

Underwriters Laboratories standards

UL 558 Internal Combustion Engine-Powered Industrial Trucks
583 Electric Battery-Powered Industrial Trucks

Rigging equipment

Forklifts

ANSI/NFPA 505 Powered Industrial Trucks, Type Designation and Areas of Use

ASME B56 Standards

B56.1-2005 Safety Standard for Low- and High-Lift Trucks (2005)
Defines the safety requirements relating to the elements of design, operation, and maintenance of low- and high-lift-powered industrial trucks controlled by a riding or walking operator, and intended for use on compacted, improved surfaces.

B56.6-2005 Safety Standard for Rough Terrain Forklift Trucks (2005)
Defines the safety requirements relating to the elements of design, operation, and maintenance

of rough terrain forklift trucks. These trucks are intended for operation on unimproved natural terrain as well as the disturbed terrain of construction sites.

B56.10-2006 Safety Standard for Manually Propelled High-Lift Industrial Trucks (2006)
Defines the safety requirements relating to the elements of design, operation, and maintenance of manually propelled high-lift industrial trucks controlled by a walking operator, and intended for use on level, improved surfaces.

B56.11.4-2005 Hook-Type Forks and Fork Carriers; Powered Industrial Forklift Trucks (2005)
Encompasses standards relative to hook-type fork carriers and the attaching elements of fork arms and load handling attachments for forklift trucks, in relation to manufacturers' rated capacities of trucks up to and including 24,000 lb (11,000 kg).

B56.11.6-2005 Evaluation of Visibility from Powered Industrial Trucks (2005)
Establishes the conditions, procedures, equipment, and acceptability criteria for evaluating visibility from powered industrial trucks. It applies to internal combustion engine-powered and electric high-lift, counterbalanced, sit-down rider industrial trucks up to and including 22,000 lb (10,000 kg) capacity, inspection, testing, and acceptance, thereby fostering universal use of export pallets in international commerce with minimal restrictions.

B56.11.7-2005 Liquefied Petroleum Gas (LPG) Fuel Cylinders (Horizontal or Vertical) Mounting—Liquid Withdrawal (2005)
Establishes dimensions for LPG fuel cylinders used on powered industrial trucks.

Hand and portable power tools
ASSE/SAFE A10.3 Safety Requirements for Powder-Actuated Fastening Systems

Helicopters
ASME B30.12 Handling Loads, Suspended from Rotorcraft

Regulations and Standards

Hoists

ASME B30.7	Base-Mounted Drum Hoists
ASME B30.23	Personnel Lifting Systems
ASME B30.21	Manually Lever-Operated Hoists
ASME/HST-1M	Performance Standard for Electric-Chain Hoists
ASME/HST-2M	Performance Standard for Hand-Chain Manually Operated Chain Hoists
ANSI A17; B30 and B56	Vertically adjustable equipment used primarily to raise and lower materials and equipment from one elevation to another
ASSE/SAFE A10.22	Hoists—Safety Requirements for Rope-Guided and Non-Guided Workers
ASSE/SAFE A10.4	Personnel Hoists and Employee Elevators on Construction and Demolition Sites
ASSE/SAFE A10.5	Safety Requirements for Material Hoists

Lifting and hoisting machinery and systems

ANSI MH27.1	Specifications for Underhung Cranes and Monorail Systems
ASME B30.2	Overhead and Gantry Cranes (Top Running Bridge, Single or Multiple Girder, Top Running Trolley Hoist)
B30.5	Mobile and Locomotive Cranes
B30.9	Slings
B30.10	Hooks
B30.11	Monorails and Underhung Cranes
B30.16	Overhead Hoists (Underhung)
B30.17	Overhead and Gantry Cranes (Top Running Bridge, Single Girder, Underhung Hoist)
B30.20	Below-the-Hook Lifting Devices
B30.21	Manually Lever-Operated Hoists
B30.23	Personnel Lifting Systems
C B30.22	Articulating Boom Cranes
C NOG-1	Rules for Construction of Overhead and Gantry Cranes (Top Running Bridge, Multiple Girder)

C NUM-1	Rules for Construction of Cranes, Monorails, and Hoist (With Bridge or Trolley or Hoist of the Underhung Type)
ASSE/SAFE	Crane Hazards and Their Prevention—50 Categories of Crane Hazards, Including Power Line Contact, Types of Upset and Pinch Points and Nip Points
A10.28	Safety Requirements for Work Platforms Suspended from Cranes or Derricks
A10.42	Safety Requirements for Rigging Qualifications and Responsibilities
Z15.1	Safe Practices for Motor Vehicle Operations
CMAA No. 70	Specification for Top Running Bridge and Gantry-Type Multiple Girder (Electric Overhead Traveling Cranes)
No. 74	Specification for Top Running and Under Running Single Girder (Electric Overhead Traveling Cranes Utilizing Under Running Trolley Hoist)
NFPA 70 Article 610	Cranes and Hoists—National Electrical Code
PCSA Std. No. 1	Mobile Power Crane (and Excavator) and Hydraulic Crane Standards
No. 2	Mobile Hydraulic Crane Standard
No. 4	Mobile Power Crane (and Excavator) and Hydraulic Crane Standards
SAE J376-95	Load-Indicating Devices in Lifting Crane Service
J765	Code: Crane Load Stability Test
J874	Code: Center of Gravity Test

Masonry and concrete work

ASSE/SAFE A10.9	Safety Requirements for Masonry and Concrete Work

Personal protection equipment

ANSI Z87.1-2003	Occupational Personal Eye and Face Protection
Z89.1-2003	Protective Headwear for Industrial Workers

B11.TR5-2006	Sound Level Measurement Guidelines for Measuring, Evaluating, Documenting, and Reporting Sound Levels Emitted by Machinery
ASSE/SAFE A10.46	Hearing Loss Prevention for Construction and Demolition Workers

Personal fall arrest systems and nets

ASSE/SAFE FALL PROTECTION	*Introduction to Fall Protection*, 3rd Edition, Identification of Specific Walking Working Surface Hazards, Including Slips and Trips, Stairways and Ramps, Ladders, Scaffolds, and Roofs
A10.6	Demolition Operations—Safety Requirements
Z359.1	Safety Requirements for Personal Fall Arrest Systems, Subsystems, and Components
Z359.2	Minimum Requirements for a Comprehensive Managed Fall Protection Program
Z359.3	Safety Requirements for Positioning and Travel Restraint Systems
Z359.4	Safety Requirements for Assisted-Rescue and Self-Rescue Systems, Subsystems, and Components
Z359	Fall Protection Code Package (Includes Z359.0, Z359.1, Z359.2, Z359.3, and Z359.4)
A10.24	Roofing—Safety Requirements for Low-Sloped Roofs
A10.32	Fall Protection Systems

Safety and debris nets

ASSE/SAFE A10.11	Safety Requirements for Personnel and Debris Nets
A10.37	Debris Net Systems Used during Construction and Demolition Operations

Portable space heating devices

ASSE/SAFE A10.10 (R 2004)	Safety Requirements for Temporary and Portable Space Heating Devices
	American National Standard for Construction and Demolition Operations

Rigging hardware

ANSI/ASME B18.15	Forged Eyebolt
ASTM C A391/A391M	Standard Specification for Grade 80 for Alloy Steel Chain
C A489	Standard Specification for Carbon Steel Eyebolts
C E165	Standard Practice for Liquid Penetrant Inspection Method
C E709	Standard Practice for Magnetic Particle Examination
C F1145	Standard Specification for Turnbuckles, Swaged, Welded, Forged
F541	Standard Specifications for Alloy Steel Eyebolts

Roof and floor holes and wall openings

ASSE/SAFE A10.18	Safety Requirements for Temporary Roof and Floor Holes, Wall Openings, Stairways, and Other Unprotected Edges in Construction and Demolition Operations
ANSI A10.18	Floor and Wall Openings, Railings, and Toe Boards

Roofing

ASSE/SAFE A10.24	Roofing—Safety Requirements for Low-Sloped Roofs

Scaffolds and work platforms

ANSI/SIA A92	Elevating and Vehicle Lift Devices Package
A92.2-2009	Vehicle-Mounted Elevating and Rotating Aerial Devices
A92.3-2006	Manually Propelled Elevating Aerial Platforms
A92.5-2006	Boom-Supported Elevating Work Platforms
A92.6-2006	Self-Propelled Elevating Work Platforms
A92.8-2006	Vehicle-Mounted Bridge Inspection and Maintenance Devices
A92.9-1993	Mast-Climbing Work Platforms

A92.10-2009 — Transport Platforms
A120.1-92 — Suspended Powered Platforms for Exterior Building Maintenance
ASSE/SAFE A10.28 — Safety Requirements for Work Platforms Suspended from Cranes or Derricks
SSFI/SIA — Scaffold Code of Safe Practices

Slings, hooks, and attachments
ASME B30.9 — Slings
B30.21 — Hooks
B30.20 — Below-the-Hook Lifting Devices

Steel erection
ASSE/SAFE A10.13 — Safety Requirements for Steel Erection

Walking/working surfaces
ASSE/SAFE
A1264.1 — Safety Requirements for Workplace Walking/Working Surfaces and Their Access; Workplace, Floor, Wall and Roof Openings; Stairs and Guardrails Systems
A1264.2 — Provision of Slip Resistance on Walking/Working Surfaces

Welding
ANSI/AWS D1.1 — Structural Welding Code—Steel
D1.2 — Structural Welding Code—Aluminum
D14.1 — Specification for Welding of Industrial and Mill Cranes and Other Material Handling Equipment

Wire rope and alloy steel chain
ANSI/ASTM A391 — Alloy Steel Chain Specification

Index

Barricades, 333
  safety barricades and lights, 333

Cranes and derricks, 49
  environmental controls, 16
  facilities, 15
  guidelines (industry), 13
  inspections, 64
    load-rating chart, 69
    maintenance, 66
    regulations, 77
      initial inspection, 77
      regular inspections, 77
  operational procedures, 49
    crawler and mobile cranes, 62
    guidelines (maintenance), 61
    overhead and gantry cranes, 62
  standards (consensus), 78
Cranes, hoists, derricks, and lifting systems, 44, 79
  crawler, truck, wheel, ringer-mounted cranes, 53
  demolition operations, 13
  derricks, cranes and hoists, 10
  energy lockout/tagout, 12
  fall protection, 11
  hoists:
    personnel and materials, 11
    wire rope and chain, 11
  masonry and concrete work, 13
  motor vehicles, 11
  nets, personnel and debris, 12
  personal protective equipment and clothing, 10

Cranes, hoists, derricks, and lifting systems (*Cont.*):
  portal, tower, and pillar cranes, 57
  powder-actuated tools, 13
Crane-specific responsibilities, 21
  construction contractors and subcontractors renting the equipment, 23
  crane and hoist operators, 23
  crane manufacturers, 22
  crane rental companies, 23
  engineering safety office, 24
  project supervisors, 23
  safety engineer (staff or third party), 24
  safe lifting operations, 58
    before lifting load, 58
    during lifting load, 59
    when lowering load, 61
  safe operating requirements, 25
    designated leader for:
      critical lifts, 27
      hoisting and rigging, 28
    equipment:
      custodian, 30
      operator, 29
    hoisting and rigging project manager, 25
    maintenance operations department, 30
    project contractor/facility manager, 25
    rigger, 29

Crane-specific responsibilities, safe operating requirements (*Cont.*):
supervisor of:
hoisting and rigging operations, 26
inspection, maintenance, and repair, 27
safety, 67
braking systems, 70
cab, 70
counterweight, 71
hydraulic hoses, fittings and tubing, 69
load hooks and hook blocks, 68
main boom, jib boom, boom extensions, 68
outriggers, 69
schedules, 71
frequent (daily), 71, 73
periodic (1- to 12-month intervals), 73, 75
walk-around (start of each shift), 76
turntable/crane body, 71
wire rope, 70
safety and health program, 9
safety organization, 21
inspections, 22
maintenance, 22
training, 22
site protection, 13
construction site, 349
demolition operations, 13
derricks, cranes and hoists, 10
energy lockout/tagout, 12
fall protection, 11
masonry and concrete work, 13
motor vehicles, 11
personal protective equipment and clothing, 10
personnel and debris nets, 12
personnel and materials hoists, 11
powder-actuated tools, 13
site protection, 13
site walk about, 345
steel erection, 13
theft and vandalism, 343
unsafe site conditions, 349

Crane-specific responsibilities, site protection (*Cont.*):
walking/working surfaces, 12
welding and cutting, 13
worksite and equipment security, 347

Electrical safety, 291
managing and preventing risks, 293
recommendations, 294
safe work practices, 298
equipment and systems, 298
workers, 298
standards (consensus), 304
Erection plan, site-specific, 36

Facilities, 15
environmental controls, 16
hazardous materials, 17
noise levels, 17
piping systems, 20
sanitary facilities, 19
ventilation, 18
exit doors, 16
exits, 15
Fall protection, 188, 195
fall arrest systems, 195
guidelines (maintenance), 196
regulations, 191, 198, 201
fall system components, 198
care and use, 200
system performance criteria, 199
walking/working surfaces, 206
hardware, 224
snaps, 224
thimbles, 224
harnesses, 223
belts and rings, 223
friction buckle, 224
lanyard, 224
tongue buckle, 223
lanyards, 224
rope lanyard, 225
steel lanyard, 224
web lanyard, 225
webbing and rope lanyard damage, 225

Fall protection (*Cont.*):
  personal fall protection arrest systems, 214
    guidelines (maintenance), 215
    regulations, 216
      arrest system connectors, 220
      lifelines and lanyards, 217
      warning line systems, 221
  safety monitoring systems, 226
  shock-absorbing packs, 226
Fencing, 329
  lighting, and signage, 326
Fire:
  protection, 336
  safety, 335
First aid/medical services, 309
Flammable and combustible materials, 338
Forklifts, 80
  guidelines, 80
  industrial forklift, 84
  inspections, 82
  maintenance, 81
  operational, 80
  regulations, 85
    material handling and storage, 86
    transporting employees and materials, 85
  standards (consensus), 88

Guardrail systems, canopies, and covers, 186
  guidelines (industry), 13
    aerial lifts (platforms), 171
    cranes and derricks, 13
    environmental controls, 16
      hazardous materials, 17
      noise levels, 17
      piping systems, 20
      sanitary facilities, 19
      ventilation, 18
    facilities, 15
      exit doors, 16
      exits, 15
    site-specific erection plan, 36
    training, 36
  regulations, 186

Hazardous, 301
  electrical work, 302
  energy control, 301
  lockout/tagout, 301
  substances communications, 342
  toxic materials, 340
Hazards:
  chemical exposure, 5, 340
  electrical, 5
  equipment, 6
  fall, 7
  fire, 6
  health, 4
  indoor air quality, 4
  rigging, 6
  tripping, slipping, and falling, 5
  ventilation, 4
Health and safety, 308
  programs (major elements), 307
Helicopters, 88
  guidelines (maintenance), 89
  hand signals, 93
  inspections, 89
  recommendations (industry), 92
  regulations, 94
  standards (consensus), 94
Hoisting and lifting procedures, 33
  crane operations, 35
  fall protection, 36
  falling object protection, 35
  inspections, 37
  lift and inspection, 38
  planning and preparation, 33
  record keeping, 42
  regulations, 34
  site-specific erection plan, 36
  standards (consensus), 44
  training, 36
Hoisting equipment, 103
  electric and air-operated hoists, 106
    air-powered hoist, 107
    electric-powered hoist, 106
  hand chain-operated hoists, 108
  material and personnel hoists, 114
    material hoists, 116
    personnel hoists, 118
  portable and overhead hoists, 103

Hoisting equipment (*Cont.*):
  winches and drums, 121
    drum systems, 123
      maintenance guidelines, 123
    winches, 122
Hoisting operations, 127
  hand signals, 127
  inspections, 107
    electric or air-powered hosts, 109
    inspection schedules, 108
  maintenance guidelines, 104, 115
  regulations, 110, 116
    electric monorail hoists, 110
    overhead hoists, 111
    regulations, winches and drums, 125
  standards (consensus), 113, 118, 128

Inspections, 37
  cranes and derricks, 64
  general work environment, 40
  helicopters, 89
  hoisting and lifting procedures, 37
  initial inspection, 77
  lift and inspection, 38
  lifting equipment, 64
  regular inspection, 71
  site walk about, 345
  visual site inspection, 37

Jacks and air lifts, 94
  maintenance guidelines, 95
  regulations, 96

Ladders, stairways, and ramps, 235
  guidelines:
    maintenance, 235
    safety, 238
  inspection (safety), 239
  maintenance (ladders), 240
  regulations (ladders), 241
  standards (consensus), 262
Ladders:
  construction site ladders, 253
  fixed ladders, 243
  general industry, 241
  maintenance guidelines, 246

Ladders (*Cont.*):
  mobile ladder stands, 252
  portable ladders, 258
  portable wood ladders, 247
  portable metal ladders, 250
Ladders or stairways, 254
  fixed stairs, 264
  ramps, 268
  stairs and stairways, 266
  fixed stairways and ramps, 263
Lift-slab operations, 98
  regulations, 96
  standards (consensus), 101
Lifting equipment, 49
  cranes and derricks, 49
  crawler and mobile cranes, 62
  crawler, truck, wheel, ringer-mounted cranes, 53
  overhead and gantry cranes, 62
  portal, tower, and pillar cranes, 57
Lifting guidelines, 61
  inspections, 64
    maintenance, 66
    safety, 67
      braking systems, 70
      cab, 70
      counterweight, 71
      hydraulic hoses, fittings, and tubings, 69
      load hooks and hook blocks, 68
      main boom, jib boom, boom extension, 68
      maintenance inspections, 66
      outriggers, 69
      safety inspections, 67
      turntable/crane body, 71
      wire rope, 70
  inspection schedules, 71
    frequent (daily), 71
    periodic (1- to 12-month intervals), 73
    walk-around (start of each shift), 75
  operational procedures, 49
  regulations, 77
    initial inspections, 71
    regular inspections, 71

Lifting guidelines (*Cont.*):
  safe lifting operations, 58
    before lifting load, 58
    during lifting load, 59
    when lowering load, 61
  safe operational procedures, 52
  standards (consensus), 44, 79
Lighting, 330
  safety barricades and lights, 333
  safety lighting, 330
  security lighting, 331
  signage, 326
Lighting, minimum illuminations intensities, 327
Load-rating chart, 69

Maintenance (guidelines), 61
  cranes and derricks:
    crawler and mobile cranes, 62
    overhead and gantry cranes, 62
  helicopters, 89
  hoisting operations, 104, 115
  jacks and air lifts, 95
  lifting equipment, 61
  personal fall protection arrest systems, 215
  portable ladders, 238, 246
  safety net systems, 228
  safety organization, 22
Medical services/first aid, 309

Operating rules, 3
Operational procedures, 49
  employer posting, 41
  medical services and first aid, 43
  record keeping, 42
  safe operational procedures, 52
  safety and health program, 42

Personal protective equipment and clothing, 311
  body, hands, and feet, 317
  equipment, 311, 312
  head and face, 314
  respiratory, 317
  sanitizing equipment and clothing, 317

Personal protective equipment and clothing (*Cont.*):
  spills, 312
  vision and hearing, 315
Portable and overhead hoists, 103
Procedures, 33
  inspections, 37
    employer posting, 41
    general work environment, 40
    lift and inspection, 38
    medical service and first aid, 43
    record keeping, 42
    safety and health program, 42
  planning and preparation, 33
  regulations, 34
    crane operations, 35
    fall protection, 36
    falling object protection, 35
    hoisting and rigging, 34
    site-specific erection plan, 36
    standards (consensus), 44
    training, 36
    visual site inspection, 37
Project responsibilities, 25
  designated leader for critical lifts, 27
  designated leader for hoisting and rigging, 28
  equipment custodian, 30
  equipment operator, 29
  hoisting and rigging project manager, 25
  maintenance operations department, 30
  project contractor/facility manager, 25
  rigger, 29
  safe operating requirements, 25
  supervisor of:
    hoisting and rigging operations, 26
    inspection, maintenance, and repair, 27

Regulations, 7
  construction rigging safety, 7
  construction scaffolds, 172
  crane operations, 35

Regulations (*Cont.*):
    cranes and derricks, 78
    fall protection, 36, 196
    falling object protection, 35
    general rigging safety, 7
    helicopters, 94
    hoisting and lifting procedures, 34
    hoisting and rigging, 34
    hoisting operations, 110
    inspection:
        initial, 71
        regular, 71
    lifting equipment, 78
    personal fall protection arrest
        systems, 216
    portable ladders, 241, 247
Rigging hardware, 145
    attachments, fittings, and
        connections, 148
    guidelines (maintenance), 148
    regulations, 147
        below-the-hook lifting
            devices, 156
        end attachments, 152
        eyebolts, 149
        fittings, 152
        shackles and hooks, 153,154
        sheaves, blocks, and tackles, 155
        standards (consensus), 157
        swivel hoist rings, 151
        wire rope clips (clamps), 149
Rigging systems, 129
    regulations, 131
        alloy steel chain slings, 135
        natural fiber rope slings and
            hitches, 137
        synthetic fiber rope and web
            slings, 140
        synthetic fiber rope slings, 140
        synthetic web slings, 141
        wire rope, wire rope slings and
            meshes, 129
            maintenance guidelines, 130
            mandatory inspections, 130
            regulations, 131
            wire rope slings, 132
                metal mesh slings, 133
    standards (consensus), 144

Safe lifting operations, 58
    before lifting load, 58
    during lifting load, 59
    when lowering load, 61
Safety, health and, 308
    programs (major elements), 307
Safety net systems, 227, 232
    guidelines (maintenance), 228
    regulations, 229, 232
    standards (consensus), 232
Scaffold:
    access, 181
    setup, 159
Scaffold use, 161
    aerial lifts (platforms), 171
        guidelines (industry), 171
    bricklayers square scaffolds, 179
    construction scaffolds, 172
        regulations, 172
    fabricated frame scaffolds, 179
    horse scaffolds, 180
    ladder jack scaffolds, 180
    mobile scaffolds, 170
    stationary scaffolds, 166
    supported scaffolds, 175
    suspended scaffolds, 165, 175
    tube-and-coupler scaffolds, 177
    wood pole scaffolds, 176
Scaffolding systems, 159, 166,
        181, 184
    guidelines (maintenance), 159
    regulations, 167, 181, 186
    standards (consensus), 19
    training, 181
Security:
    construction site, 347
    equipment, 347, 348
    site conditions (unsafe), 349
    worksite and equipment, 347
Signage, 326, 332
    fencing, 326
    lighting, 326
Site, construction, 347
    unsafe conditions, 349
Spills (personal equipment), 312
Standards (consensus), 9
    barricades, 333
    cranes and derricks, 79

Standards (consensus) (*Cont.*):
  demolition operations, 13
  derricks, cranes and hoists, 10
  energy lockout/tagout, 12
  fall protection, 11
  fencing, 334
  helicopters, 94
  hoisting and lifting procedures, 44
  hoisting operations, 112, 118, 128
  hoists (wire rope and chain), 11
  lighting, 334
  masonry and concrete work, 13
  motor vehicles, 11
  personal protective equipment and clothing, 10
  personnel and debris nets, 12
  personnel and materials hoists, 11
  portable ladders, 262
  powder-actuated tools, 13
  rigging systems, 143
  safety and health program, 9
  safety net systems, 227
  signage and barricades, 334
  site protection, 13, 349
  steel erection, 13
  theft and vandalism, 343
  tools and machinery, 289
  training, 181
  walking/working surfaces, 12
  walkways and elevated platforms, 272

Tools and machinery, 273
  compressors and compressed air, 284
  gas cylinders, 285
    compressed gas, 285
    fuel gas, 286
  hand and portable power tools, 273
  hand and power tools, 278
    hand tools, 280
    powder-actuated tools, 281

Tools and machinery (*Cont.*):
  machine guarding and lockout, 287
  pneumatic power tools and hoses, 276
  regulations, 274
  standards (consensus), 289
  welding and cutting (hot work), 281

Vandalism, theft and, 343

Walking/working surfaces, 206
  dangerous equipment, 209
  excavations, 208
  guardrail systems, 211
  hoist area, 207
  holes, 208
  overhead bricklaying and related work, 209
  precast concrete erection, 210
  ramps, runways and other walkways, 208
  roofing work on low-slope roofs, 209
  steep roofs, 210
  training, 213
  unprotected sides and edges, 207
  wall openings, 210
Walkways and elevated platforms, 266
  elevated surfaces, 267
  floor and wall openings, 268
  general areas and walkways, 266
  guarding roofs, floors, stairs, and other openings, 268
  regulations, 268
  standards (consensus), 272
Worker protection, 307
  health and safety, 308
    major elements, 307
    medical services/first aid, 309